Semiparametric Regression
for the Social Sciences

Semiparametric Regression for the Social Sciences

Luke Keele

Ohio State University, U.S.A.

John Wiley & Sons, Ltd

This publication is designed to provide accurate and authoritative information in regard to the subject
matter covered. It is sold on the understanding that the Publisher is not engaged in rendering professional
services. If professional advice or other expert assistance is required, the services of a competent
professional should be sought.

Other Wiley Editorial Offices

John Wiley & Sons Inc., 111 River Street, Hoboken, NJ 07030, USA

Jossey-Bass, 989 Market Street, San Francisco, CA 94103-1741, USA

Wiley-VCH Verlag GmbH, Boschstr. 12, D-69469 Weinheim, Germany

John Wiley & Sons Australia Ltd, 42 McDougall Street, Milton, Queensland 4064, Australia

John Wiley & Sons (Asia) Pte Ltd, 2 Clementi Loop #02-01, Jin Xing Distripark, Singapore 129809

John Wiley & Sons Canada Ltd, 6045 Freemont Blvd, Mississauga, ONT, L5R 4J3

Wiley also publishes its books in a variety of electronic formats. Some content that appears in print may
not be available in electronic books.

Library of Congress Cataloging-in-Publication Data

Keele, Luke, 1974-
    Semiparametric regression for the social sciences / Luke Keele.
        p. cm.
    Includes bibliographical references and index.
    ISBN 978-0-470-31991-8 (cloth)
1. Regression analysis. 2. Nonparametric statistics. I. Title.
    QA278.2.K42 2008
    519.5'36–dc22   2007045557

British Library Cataloguing in Publication Data

A catalogue record for this book is available from the British Library

ISBN: 978-0-470-31991-8

Typeset in 10.5/13pt Times by Thomson Digital, Noida

Contents

List of Tables

List of Figures

Preface

This is a book for analysts from the social sciences who would like to to extend their statistical toolkit beyond standard parametric functional forms. In general, the methods presented in this book alter the use of standard models. That is, the techniques here are not meant to be a replacement for models such as linear or logistic regression, but instead are meant to enhance how these models are used. As a result, the techniques can enhance data analysis but they are not a panacea. While nonparametric regression can flexibly estimate nonlinear functional forms, it cannot make a correlation into a causal effect, it cannot make up for an omitted variable, and it cannot prevent data mining. The visual aspect of these models, however, can make analysts more sensitive to patterns in the data that are often obscured by simply reading a parameter estimate off the computer screen.

The goal of this book is to make applied analysts more sensitive to which functional form they adopt for a statistical model. The linear functional form is ubiquitous in social science research, and it often provides a very reasonable approximation to the process we are trying to understand. But it can also be wrong, leading to inferential errors, especially when applied without much thought as to whether it is justified. We do not presume that all relationships are nonlinear, but that the assumption of linearity should be tested, since nonparametric regression provides a simple and powerful tool for the diagnosis and modeling of nonlinearity

This book is also designed to be attuned to the concerns of researchers in the social sciences. There are many books on this topic, but they tend to focus on the concerns of researchers in the sciences and are full of examples based on the patterns of sole eggs and the reflection of light off gas particles. To that end, we provide a hands-on introduction to nonparametric and semiparametric regression techniques. The book uses a number of concrete examples, many from published research in both political science and sociology, and each example is meant to demonstrate how using nonparametric regression models can alter conclusions based on standard parametric models. The datasets and R code for every example can be found on the book webpage at: http://www.wiley. com/go/keele_semiparametric. Morever, the appendix provides a basic overview of how to produce basic results.

Since the techniques presented here alter the estimation of standard models, we assume that the reader is familiar with more standard models. For example, while we provide an example of estimating a semiparametric count model, we do not outline aspects of count models. Instead, we refer readers to more authoritative sources. In general, various chapters have differing levels of presumed knowledge. Chapters 1–3 presume a basic familiarity with linear regression models at a level that would be similar to a course in regression in a graduate first-year methods sequence. That said, various topics within those chapters may be more advanced. If this is the case, we alert readers. Chapter 6 presumes that readers are familiar with more advanced models including discrete choice models such as logit and probit, count models, and survival models. The presentation of these models, however, is self-contained such that readers familiar with logit models can read the relevant sections without needing to know anything about survival models. The more advanced topics in Chapter 7 again each presume specialized background information but are again self-contained.

I would like to thank the many people who provided feedback and helpful comments including Chris Zorn, Alan Wiseman, Irfan Nooruddin, Marcus Kurtz, Charles Franklin, Karen Long Jusko, Michael Colaresi, Mike Neblo, Craig Volden, Jan Box-Steffensmeier, Neal Beck, Jas Sekhon, Anand Sokhey, Ryan Kennedy, Corrine McConnaughy, Mark Kayser, and the many anonymous reviewers. I am also indebted to all the people who provided me with data for use in the illustrations, and to my students for allowing me to try the material out on them. I also want to acknowledge Jim Stimson and Walter Mebane for general guidance and advice on what it means to do careful research using data. Finally, I would like to thank my wife Tracey for not begruding all the nights and weekends spent with data instead of her. The author would appreciate being informed of any errors and may be contacted at: keele.4@osu.edu.

Luke Keele

1

Introduction: Global versus Local Statistics

Statistical models are always simplifications, and even the most complicated model will be a pale imitation of reality. Given this fact, it might seem a futile effort to estimate statistical models, but George Box succinctly described the nature of statistical research: 'All models are wrong, some are useful.' Despite the fact that our models are always wrong, statistics provides us with considerable insight into the political, economic and sociological world that we inhabit. Statistical models become simplifications of reality because we must make assumptions about various aspects of reality. The practice of statistics is not shy about the need to make assumptions. A large part of statistical modeling is performing diagnostic tests to ensure that the assumptions of the model are satisfied. And in the social sciences, a great deal of time is devoted to checking assumptions about the nature of the error term: are the errors heteroskedastic or serially correlated? Social scientists are far more lax, however, when it comes to testing assumptions about the functional form of the model. In the social sciences, the linear functional (and usually additive) form reigns supreme, and researchers often do little to verify the linearity assumption. Much effort is devoted to specification and avoiding misspecification, but little is done to explore other functional forms when the incorrect functional form is in essence a specification error.

Semiparametric Regression for the Social Sciences   Luke Keele
© 2008 John Wiley & Sons, Ltd

The reliance on linear functional forms is more widespread than many probably realize. For many analysts, the question of linearity is focused solely on the nature of the outcome variable in a statistical model. If the outcome variable is continuous, we can usually estimate a linear regression model using least squares. For discrete outcomes, analysts typically estimate generalized regression models such a logistic or Poisson regression. Researchers often believe that because they are estimating a logistic or Poisson regression model, they have abandoned the linear functional form. Often this is not the case, as the functional form for these models remains linear in an important way. The generalized linear model (GLM) notation developed by McCullagh and Nelder (1989) helps clarify the linearity assumption in models that many researchers think of as nonlinear.

In the GLM framework, the analyst makes three choices to specify the statistical model. First, the analyst chooses the stochastic component of the model by selecting a sampling distribution for the dependent variable. For example, we might choose the following sampling distribution for a continuous outcome:

$$Y_i \sim N(\mu_i, \sigma^2). \tag{1.1}$$

Here, the outcome Y_i follows a Normal sampling distribution with expected value μ_i and a constant variance of σ^2. This forms the stochastic component of the statistical model. Next, the analyst must define the systematic part of the model by choosing a set of predictor variables and a functional form. If we have data on k predictors, $\mathbf{X}$ is an $n \times k$ matrix containing k predictors for n observations, and η is an $n \times 1$ vector of linear predictions:

$$\eta = \mathbf{X}'\boldsymbol{\beta} \tag{1.2}$$

where $\boldsymbol{\beta}$ is a vector of parameters whose values are unknown and must be estimated. Both $\mathbf{X}'\boldsymbol{\beta}$ and η are interchangeably referred to as the linear predictor. The linear predictor forms the systematic component of the model. The systematic component need not have a linear functional form, but linearity is typically assumed. Finally, the analyst must choose a *link* function. The link can be as simple as the identity function:

$$\mu_i = \eta. \tag{1.3}$$

The link function defines the connection between the stochastic and systematic parts of the model. To make the notation more general, we can write the link function in the following form:

$$\mu_i = g(\eta_i), \tag{1.4}$$

where $g(\cdot)$ is a link function that must be monotonic and differentiable. When the stochastic component follows a Normal distribution and the identity function links the stochastic and systematic components, this notation describes a linear regression model. The GLM framework, however, generalizes beyond linear regression models. The stochastic component may come from any of the exponential family of distributions, and any link function that is monotonic and differentiable is acceptable.

As an example of this generality, consider the GLM notation for a model with a binary dependent variable. Let Y_i be a binary variable without outcomes 0/1, where 1 represents a success for each m_i trials and φ_i is the probability of a success for each trial. If this is true, we might assume that Y_i follows a Bernoulli distribution, which would imply the following stochastic component for the model:

$$Y_i \mid \varphi_i \sim B(m_i, \varphi_i). \tag{1.5}$$

The systematic component remains $\mathbf{X}'\boldsymbol{\beta}$, the linear predictor. We must now select a link function that will ensure that the predictions from the systematic component lie between 0 and 1. The logistic link function is a common choice

$$\varphi_i = \frac{1}{1 + e^{-\eta_i}}. \tag{1.6}$$

With the logit link function, no matter what value η_i takes the predicted value for Y_i, φ_i, will always be between zero and one.

What has been the purpose of recounting the GLM notation? The point of this exercise is to demonstrate that while the link function is a nonlinear transformation of the linear predictor, the systematic component of the model remains *linear*. Across the two examples, both the link function and stochastic component of the model differed, but in both cases we used the linear predictor $\mathbf{X}'\boldsymbol{\beta}$. The second model, a logistic regression, is thought of as a nonlinear model, but it has a linear functional form. Thus many of the models that analysts assume are nonlinear models retain a linearity assumption. There is nothing about the logistic regression model (or any other GLM) that precludes the possibility that the model is nonlinear in the variables. That is, instead of the effect of X on Y, which is summarized by the estimated β coefficient, being constant, that effect varies across the values of X. For example, one is as likely to use a quadratic term in a Poisson regression model as a linear regression model.

Why, then, is the assumption of linearity so widespread? And why is model checking for deviations from nonlinearity so rare? This question cannot be answered definitively. Quite possibly, the answer lies in the nature of social science theory. As Beck and Jackman (1998) note: 'few social scientific theories offer any guidance as to functional form whatsoever.' When researchers stipulate the predicted relationship between X and Y, they do not go beyond 'the relationship between X and Y is expected to be positive (or negative).' Given that most theory is silent as to the exact functional form, linearity has become a default, while other options are rarely explored. We might ask what are the consequences of ignoring nonlinearity?

1.1   The Consequences of Ignoring Nonlinearity

What are the consequences for using the wrong functional form? Simulation provides some useful insights into how misspecification of functional form can affect our inferences. First, we need some basic notation. We start with the linear and additive functional form for Y

$$Y = \beta_0 + \beta_1 X_1 + \beta_2 X_2. \tag{1.7}$$

The above equation is linear in the parameters in that we have specified that the mean of Y is a linear function of the variables X_1 and X_2. It is possible that the effect of X_2 on Y is nonlinear. If so, the effect of X_2 on Y will vary across the values of X_2. Such a model is said to be nonlinear in the variables but linear in the parameters. Estimation of models that are nonlinear in the variables presents few problems, which is less true of models that are nonlinear in the parameters. Consider a model of this type

$$Y = \beta_0 + \beta_1 X_1^{\beta_2}.$$

Here, the model is nonlinear in the parameters and estimation is less easy as nonlinear least squares is required. For a model that is nonlinear in the variables, we can still use least squares to estimate the model. What are the consequences of ignoring nonlinearity in the variables if it is present? For example, assume that the true data generating process is

$$Y = \beta_0 + \beta_1 X_1 + \beta_2 X_1^2. \tag{1.8}$$

Suppose an analyst omits the quadratic term when estimating the model. The effect of omitting the nonlinear term from the right hand side of a regression model is easily captured with simulated data. To capture the above data generating process, we simulate X_1 as 500 draws from a uniform distribution on the interval 1 to 50. Y is a function of this X variable with an error term drawn from a Normal distribution with zero mean and constant variance and the β parameters are set to ones. What are the consequences of fitting a linear model without the quadratic term? Figure 1.1 contains a plot of the estimated regression line when only a linear term has been included on the right hand side of the estimated model and the true quadratic functional form.

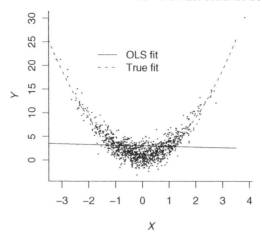

Figure 1.1   A linear fit and the true quadratic relationship with simulated data.

Here, the model with the incorrect functional form would lead one to conclude that X is unrelated to Y as the regression line is virtually flat, when in fact the curved line reflects the strong but nonlinear relationship between the two variables. This simulated example illustrates the misspecification that results from assuming linearity when the relationship between X and Y is actually nonlinear. In this particular example, the consequences are particularly severe, given that an analyst might conclude that there is no relationship between X and Y, when the two are strongly related. Moreover, as the GLM framework implies, models such as logistic or Poisson regression are equally prone to this misspecification from an incorrect functional form. If X^2 is in the true data generating process, no link function will correct the specification error. In these models, the systematic component of the model often remains linear and the failure to include a nonlinear term in the model will have equally deleterious effects on the model estimates. In fact, given the serious distortions such a misspecification can cause, it would be prudent to test that the effect of any continuous covariate on the right hand side of a model does not have a nonlinear effect. If we admit that defaulting to linearity might be a problematic method of data analysis, what alternatives are available? While there are a number of possibilities, analysts in the social sciences usually rely on power transformations to address nonlinearity.

1.2   Power Transformations

Power transformations are a simple and flexible means of estimating a nonlinear functional form. While a variety of power transformations are possible, most researchers restrict themselves to only one or two transformations. A common notation exists for all power transformations that is useful to outline since we will often use power transformations in the data analytic examples that follow. For a strictly positive variable, X, we can define the following set of power transformations:

$$X^\lambda. \tag{1.9}$$

Using different values for λ produces a wide variety of transformations. If λ is 2 or 3, the transformation is quadratic or cubic respectively, and if λ is $1/2$ or $1/3$, the transformation is either the squared or cubic root. By convention, a value of 0 for λ denotes a log transformation, and a value of 1 for λ corresponds to no transformation (Weisberg 2005).

Power transformations are often a reasonable method for modeling nonlinear functional forms. For example, it is well understood that older people vote more often than younger people (Nagler 1991). So as age increases, we expect that the probability of a person voting increases. But we do not expect the effect of age on voter turnout to be linear since once people reach a more advanced age, it often prevents them from voting. To capture such nonlinearity, most analysts rely on a quadratic power transformation. In this model, the analyst would square the age variable and then include both the untransformed and squared variable on the right hand side of the model. If the quadratic term is statistically significant at conventional levels ($p < 0.05$), the analyst concludes that the relationship between the two variables is nonlinear.

The process described above is often a reasonable way to proceed. Some care must be taken with the interpretation of the quadratic term as the standard error for the marginal effect must be calculated by the analyst, but the transformation method is easily to use. Moreover, the model is now a better representation of the theory, and power transformations avoid any misspecification due to an incorrect functional form. If the effect of age on voter turnout is truly quadratic, and we only include a linear term for age; we have misspecified the model and biased not only the estimate of age on voter turnout but all the other estimates in the model as well.[1]

Power transformations, however, have several serious limitations. First and foremost, power transformations are global and not local fits. With a global fit, one assumes that the statistical relationship between X and Y does not vary over the range of X. When an analyst estimates a linear relationship or uses a

---

[1]This is true since I assume the analyst has estimated a model such as a logit. In linear regression, misspecification only biases the coefficient estimates if the omitted variable is correlated with other variables, but for most models with nonlinear link functions, all the parameters will be biased if the model is misspecificed.

power transformation, the assumption is that the relationship between X and Y is exactly the same for all possible values of X. The analyst must be willing to assume a global relationship when he or she estimates linear functional forms or uses power transformations. Quite often the relationship between X and Y is local: the statistical relationship between two variables is often specific to local regions of X. The assumption of a global relationship often undermines power transformation.

For example, quadratic power transformations assume that the relationship between X and Y for all values of X is strictly quadratic regardless of whether this is true or not. When it is not, the power transformation can overcorrect the nonlinearity between X and Y. While there are a variety of power transformations, each assumes a global form of nonlinearity, which may or may not be correct. A variety of more complex nonlinear forms cannot be easily modeled with power transformations. Therefore, while power transformations can model some forms of nonlinearity, the global nature of power transformation fits often undermines them as a data analytic technique. In several of the data examples that follow, we see how power transformations often cannot adequately capture the nonlinear relationship in the data.

More seriously, the choice about which power transformation to use is often arbitrary. At best our theory might indicate that an effect is nonlinear, such as it does with age and voter turnout, but to have a theory that actually indicates that the effect is quadratic is unlikely. For example, how might one choose between a logarithmic or quadratic power transformation? One can estimate a model where the value of λ is estimated as the value that minimizes the sum of squared model errors. Unfortunately, this method often produces transformations that are uninterpretable (Berk 2006).

Moreover, the choice of which power transformation to use is not without consequences. For example, one might suspect a nonlinear relationship between X and Y and both a logarithmic and quadratic transformation seem like reasonable corrections. Does it matter which one the analyst uses? Another example with simulated data brings the problem into sharper focus. The simulated Y variable is a logarithmic function of a single X variable that is a drawn from a uniform distribution on the interval one to 50. The data generating process for Y is

$$Y_i = 0.50 \log (X_i) + \varepsilon \qquad (1.10)$$

where ε is distributed IID Normal and the sample size is 500. We expect the estimate for β in the model to be 0.50. To better understand the consequences of an incorrect functional form, we estimate three different regression models with differing functional forms. The first model has the correct functional form with a logged X, the second model uses a squared X, and the last model has a linear functional form. The estimate from the model with logged X is 0.497 – very close to the true parameter value. The estimates from the models with the incorrect functional forms are 0.030 for the linear model and 0.049 for the quadratic, well under the true parameter value. The consequences of both ignoring the nonlinearity and using the wrong transformation are abundantly clear. For both the quadratic and linear fits, the parameter estimate is highly attenuated, so how might we choose the correct functional form? The model fit is not indicative of the wrong functional form as the R^2 values across the three models are: 0.168, 0.168, and 0.171. Perhaps a measure of nonnested model fit will provide some evidence of which model is to be preferred? The Akaike Information Criterion (AIC) is a measure of nonnested model fit estimated by many statistical software packages (Akaike 1973). Lower AIC values indicate better model fit to the data. The AIC values for the logarithmic, linear, and quadratic fits are respectively: 1409.35, 1409.47, and 1409.04, so the AIC points to the incorrect quadratic transformation. If the nonlinearity were quadratic instead of logarithmic, we could expect a similar quandary. This simulation demonstrates that the choice of power transformation is both important and difficult. The wrong power transformation can seriously affect the model estimates, but the analyst has few tools to discriminate between which transformation to use.

Moreover, since power transformations can only be used with strictly positive X variables, the analyst is often forced to arbitrarily add positive constants to the variable. Finally, power transformations are only effective when the ratio of the largest values of X to the smallest values is sufficiently large (Weisberg 2005).

In sum, power transformations are useful, but they have serious drawbacks and as such cannot be an analyst's only tool for modeling nonlinearity in regression models. What other alternatives are available? The most extreme option is to use one of several computationally intensive methods such as neural networks, support vector machines, projection pursuit, or tree based methods (Hastie, Tibshirani, and Friedman 2003). The results from such techniques, however, can be hard to interpret, and it can be difficult to know when the data are overfit. A less extreme alternative is to use nonparametric and semiparametric regression techniques.

1.3   Nonparametric and Semiparametric Techniques

Instead of assuming that we know the functional form for a regression model, a better alternative is to estimate the appropriate functional form from the data. In the absence of strong theory for the functional form, this is often the best way to proceed. To estimate the functional form from data, we must replace global estimates with *local* estimates. With global estimates, the analyst assumes a functional form for the model; with local estimates, the functional form is estimated from the data. The local estimators used in this text are referred to as nonparametric regression models or smoothers. Nonparametric regression allows one to estimate nonlinear fits between continuous variables with few assumptions about the functional form of the nonlinearity. Both *lowess* and splines are common nonparametric regressions models that rely on local estimates to estimate functional forms from data.

What is local estimation? With local estimation, the statistical dependency between two variables is described not with a single parameter such as a mean or a β coefficient, but with a series of local estimates. For local estimators, a estimate such as a mean or regression is estimated between Y and X for some restricted range of X and Y. This local estimate of the dependency between X and Y is repeated across the range of X and Y. This series of local estimates is then aggregated with a line drawing to summarize the relationship between the two variables. This resulting nonparametric estimate, which may be linear or quadratic, does not impose a particular functional form on the relationship between X and Y. Due to the local nature of the estimation process, nonparametric regression provides very flexible fits between X and Y. Local models such as nonparametric regression estimate the functional form between two variables while global models impose a functional form on the data.

Nonparametric regression models are useful for both the diagnosis of nonlinearity and modeling nonlinear relationships. A plot of the estimated residuals against the fitted values is a standard diagnostic for most statistical models. Trends in the variability of the residuals against the fitted values is, for example, a sign of heteroskedasticity. When examined in conjunction with nonparametric regression such plots are far more informative for the diagnosis of nonlinear functional forms. Any trend in the nonparametric regression estimate is a sign of various forms of misspecification including unmodeled nonlinearity. Figure 1.2 contains a plot of the residuals against the fitted values for the first example with simulated data. In that model, the true functional form was quadratic but we fit a linear model. The specification error is readily apparent in this plot. The plot also contains the estimate from a nonparametric regression model. The nonparametric regression estimate closely matches the quadratic functional form without the analyst having to assume he or she knows the true functional form. While such plots are useful diagnostics, a plot of this type can only tell us there is a problem. What is needed is a multivariate model that allows one to combine global and local estimates. The solution is the semiparametric regression model.

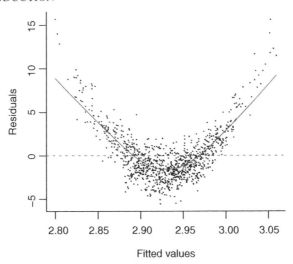

Figure 1.2   Residuals plotted against fitted values with a nonparametric regression estimate.

Semiparametric regression models are often referred to as either additive or generalized additive models (GAMs). Additive models and GAMs incorporate local estimation models like *lowess* and splines into standard linear models and GLMs. Given that the additive model is a special case of a GAM, throughout the text the term GAM will be used to refer to semiparametric regression models for both continuous and discrete outcome variables. These models allow the analyst to model some predictor variables with nonparametric regression, while other predictor variables are estimated in a standard fashion. A single continuous X variable suspected of having a nonlinear functional form may be modeled nonparametrically, while the rest of the model specification is estimated parametrically. Given that GAMs rely on nonparametric regression, the assumption of a global fit between X and Y is replaced with local fitting. While GAMs relax the assumption of a global fit, they do not dispense with the assumption of additive effects. The additivity assumption of GAMs makes the models easier to interpret than neural networks, support vector machines, projection pursuit, and tree based methods, but more flexible than a fully parametric model. The GAM framework is easily adapted to standard social science data analysis; GAMs work with a variety of dependent variables: interval, count, binary, ordinal, and time-to-event.

Most importantly, GAMs provide a framework for the diagnosis of nonlinearity. The simple linear and power transformation fits are nested within GAMs.

Therefore, the local estimates from a GAM can be tested against a linear, quadratic or any other transformation using either an F-test or a likelihood ratio test. If the semiparametric fit is superior to either a linear or quadratic fit, it should be used. If there is little difference between the local fit and the global fit, one can proceed with the global estimate. It is this ability to test for nonlinearity and if necessary model it that gives semiparametric regression models their power. In short, every analysis with continuous (or semicontinous) X variables requires the use of semiparametric regression model for diagnostic purposes.

At this point, one might ask if there are any objections to GAMs, and why are they not used more widely in the social sciences? A common objection is that GAMs are computationally intensive. In statistics, GAMs have often been classified as a computationally intensive technique, and semiparametric regression models are more computationally intensive than standard models since each nonparametric regression is the result of multiple local fits. The speed of modern computers has drastically reduced the time required for estimation. While a GAM may take longer than a standard parametric model, the difference in time is usually no more than a few seconds. Even with very large data sets, modern computers can now estimate most any GAM in less than 30 seconds.[2]

Some researchers object that the results from GAMs are more difficult to interpret than the results from standards models. This objection arises from the fact that GAMs do not produce a single parameter that summarizes the relationship between X and Y, but instead produces a plot of the estimated relationship between the two variables. Given that the estimation process is local as opposed to global, it is impossible to use a single parameter to describe the statistical relationship. While the absence of parameters does not allow for precise interpretation, a plot often does a far better job of summarizing a statistical relationship. With a plot, the scale of both variables should be obvious, and the eye can immediately discern the strength of the relationship along with the level of statistical precision so long as confidence bands are included.

It is often charged that GAMs overfit the data and produce highly nonlinear estimates that analysts are prone to overinterpret. It is true that an analyst may choose to undersmooth the nonparametric estimate, which can produce highly idiosyncratic and nonlinear fits between variables. Such overfitting, however, is the result of poor statistical technique; any method can be abused, and GAMs are no exceptions. Moreover, newer nonparametric regression models rely on penalized estimation which makes it more difficult to overfit the data.

---

[2]It should be noted that with very large data sets, those larger than $10,000$ cases, a GAM can take much longer to estimate than a standard model

Finally, a caveat. GAMs are not a panacea. They do not make up for poor theory or garbage can specifications. They cannot provide estimates of causal effects for observational data. The GAM is another model that should be in the toolkit of anyone that analyzes social science data. Like any statistical model, they must be used with care. Used correctly, GAM are a powerful tool for modeling nonlinear relationships. Used incorrectly, they can provide nonsensical results just as any model can.

1.4   Outline of the Text

Generalized additive models are inextricably linked to nonparametric regression, and thus one cannot understand semiparametric regression or competently estimate these models without a solid grasp of nonparametric regression. Consequently, Chapters 2–4 are devoted to various types of nonparametric smoothing techniques. These smoothers have a number of uses in their own right including exploring trends in scatterplots. As such, they are often used as a preliminary diagnostic to semiparametic regression. These chapters can be skipped if the reader is already familiar with nonparametric regression models.

The next two chapters explore how these nonparametric techniques can be married to the standard models that are used in the majority of quantitative social science research. While technical material is presented in these chapters, examples with real data from the social sciences are a point of emphasis. The next chapter explores the use of nonparametric smoothers in conjunction with mixed models, Bayesian estimation, and the estimation of propensity scores. The final chapter outlines the bootstrap – a nonparametric resampling method for estimating statistical precision – with special applications to semiparametric regression models. While bootstrapping is useful in its own right, it is critical for use with some types of semiparametric regression models. The appendix provides coverage of the software used for the analyses in the book.

The pedagogical philosophy behind this book is that new techniques are best learned through example. As such, each technique is explored with real data. In fact, the best argument for GAMs are examples of GAMs used with social science data. The majority of the examples are from published sources to demonstrate how semiparametric regression can alter the inferences made in research. All of the data sets and the computer code used in the analyses are available on the web at http://www.wiley.com/go/keele_semiparametric. There are also exercises at the end of each chapter. Many of these exercises require data analysis with applications from published research, so researchers can learn to use these models in their own work.

2

Smoothing and Local Regression

Generalized additive models are built on the foundation of smoothing and nonparametric regression techniques, and to competently use GAMs, one needs to be familiar with the principles of smoothing and nonparametric regression. As such, a thorough introduction to nonparametric regression is needed before we can turn to semiparametric regressions models. To that end, this chapter and the next cover a variety of smoothing and nonparametric techniques. While not all forms of smoothing and nonparametric regression outlined in this text are used with semiparametric regression models, one needs to understand these techniques and their precursors to be a knowledgeable user of nonparametric regression. Many of the very simple nonparametric regression models share the same basic principles as the more complex models, so for pedagogical reasons it makes sense to understand these simpler but obsolete techniques. Moreover, understanding how to use nonparametric regression is useful in its own right for both diagnostics and exploratory data analysis. This chapter starts with the simplest forms of smoothing and moves on to the widely used local polynomial regression models known as *loess* and *lowess*.

Semiparametric Regression for the Social Sciences   Luke Keele
© 2008 John Wiley & Sons, Ltd

2.1   Simple Smoothing

Smoothing and nonparametric regression are generally interchangeable terms for a set of statistical techniques used to summarize bivariate relationships in scatterplots. With more commonly used 'parametric' statistical techniques, the relationship between two variables, x and y, is summarized with a parameter such as a regression or correlation coefficient. With nonparametric regression, there is no single parameter; instead, the statistical relationship between x and y is summarized with a line drawing. Given that there is no single parameter produced by the the statistical model, people often refer to these models as nonparametric. A more formal comparison of the parametric and nonparametric approaches is useful.

Assume that we have y and x, two continuous variables, and we wish to estimate the mean of y conditional on the regressor x. We can write this relationship formally as:

$$y_i|x = f(x_i) + \varepsilon_i. \tag{2.1}$$

The term f refers to the functional form of the relationship between y and x, and f is some population function that we only assume is smooth.[1] The familiar linear functional form is a special case of this family of functional forms where the following is true:

$$f = \alpha + \beta_1. \tag{2.2}$$

Since a linear relationship is a smooth function, it is a special case of the family of smooth functions that comprise f. The linear regression model is parametric because the parameter, β_1, summarizes the statistical dependency between x and y. If we use a nonparametric regression model, there will not be a single parameter, but we make fewer assumptions. In the linear regression model, we assume that f is known to be linear, and we proceed to estimate $\hat{\beta}$. While it is possible that our estimate of f will be linear, it is also possible that f will be some nonlinear function. With nonparametric regression, we estimate the functional form for f from the data. Therefore, the *a priori* assumption of linearity is replaced with the much weaker assumption of a smooth population function. The costs of this weaker assumption are twofold. First, the computational costs are greater, though given the speed of modern computing this is no longer a significant cost. Second, we lose some interpretability, but this loss of interpretability may be worth it, since we gain a more representative estimate of the regression

---

[1] A more formal definition of smooth would be we assume that the first derivative is defined at all points of the function.

function. There are, as the reader will see, a variety of methods for the nonparametric estimation of f, and we start with the simplest of these methods: local averaging.

2.1.1 Local Averaging

The method of local averaging is best learned through a simple example. Perhaps we are interested in the statistical relationship between age and income, and we suspect that income increases with age until people retire when it then levels off or declines. The parametric approach to this problem would be to estimate a linear regression model or correlation coefficient both of which assume a linear functional form. Instead, we can summarize the relationship between age and income using a series of local averages. Here, we calculate the average income, $\bar{y}$, for each value of x, that is for each age level which would result in an estimate of $\bar{y}$ for each level of age. To form the local average estimate, we plot the average income at each age level and connect each local average with a line segment. The local average estimate is nonparametric in the sense that no single average describes the dependency between age and income, but a series of local averages become the nonparametric estimate. The resulting plot is then examined to infer the relationship between x and y. If the resulting line plot is linear and strongly sloped, we conclude that there is strong linear relationship between the two variables, but the local average estimate allows the overall estimate to vary across the level of age. The nonparametric estimate then can also be nonlinear. Importantly, since the nonparametric estimate is based on the local estimates of $\bar{y}$, it makes no assumption about the functional form of the relationship between these two variables. In fact, for this estimator we do not even assume that the underlying function is smooth.

Taking local averages at each value of x is the simplest nonparametric regression model that we can estimate. This nonparametric regression model is called a moving average smoother. The great advantage of the moving average is its simplicity, but using local averages for each value of x has serious drawbacks. Even with a large sample, we may not have many observations of income for each age, and the local estimates of $\bar{y}$ will be imprecise and fluctuate wildly across the values of x. We can decrease the variability in the local estimate of $\bar{y}$ by using a different method for dividing the range of x for the local estimates. Instead of using each value of x for the local estimate, we can divide the range of x into intervals or 'bins'. For example, the local averages could be based on age intervals of five year. So long as each each bin contains enough data points, the local estimates of $\bar{y}$ will be stable. Importantly, the variability of the local estimates of $\bar{y}$ is directly affected by the width of the bins.

Narrowing the bins reduces the number of data points used for each local estimate of $\bar{y}$, and this increases the variability across the local estimates. Increasing the bin width increases the number of data points in each bin and decreases the variability across the local estimates of $\bar{y}$. This relationship between bin width and the variability in the local estimates has a direct effect on the nonparametric estimate formed from the line plot between local averages. Wider bin widths result in a smoother fit between points, while narrow bin widths causes the line between the local estimates to be highly variable. Selecting the correct bin width is an important concept in nonparametric regression, and one that we will revisit frequently in this chapter. With large data sets, there is not much cost to having narrow bins as there will always be plenty of observations in each bin. For many data sets, however, overly narrow bin widths will provide a nonparametric estimate that is too rough.

There are several different method for 'binning' the data. One method is to divide x into bins of equal width, but the bin averages will be stable only if the data are uniformly distributed across the range of x. While this is possible, usually we cannot expect the x variable to follow a uniform distribution. A second option is to divide the range of x up into bins containing roughly equal numbers of observations, but in practice, this method will still produce nonparametric estimates that are highly variable without many observations. The solution is to avoid dividing the range of x into non-overlapping intervals. Instead, we define a moving window bin that moves across the range of x, and observations move in and out of the bin. As such, observations are used repeatedly in the calculation of the local averages that make up the nonparametric estimate. This type of bin uses the focal point x_0, the value of x that falls exactly in the middle of each bin, to define the width. Now we select the width of the bin by deciding how many x values to include on either side of the focal point.[2] To ensure that the focal point falls exactly in the middle of each bin, it is useful to select an odd number of observations for the bin width. For example, if the bin width is 31 observations, x_0, the focal point would be observation x_{16} with the 15 observations on each side of the focal point making up the complete bin width. The local average for this range of x values is the local estimate for $\bar{y}$. To move the bin across the range of x, observation x_{17} becomes the focal point of the next bin, and we take the average across the 30 values of x on either side of this new focal point. As before, expanding the bin width causes the local average to be computed across a higher number of observations and decreases the variability in the nonparametric

---

[2]The notation that we use here is conventional in many texts, but is counterintuitive since it does not respect the ordering of the x values.

estimate formed from the local averages. One drawback to the moving window bin is that it does not produce good estimates for the minimum and maximum values of x. When x_1 is the first focal point, there won't be any observations for one half of the bin, with the same being true for x_n, where n is the sample size. This causes an artificial flattening of the line called a 'boundary bias'. At this point, a concrete example is useful to help the reader understand the moving average estimator.

Example: Congressional Elections

In this chapter and the next, we use a single example data set. This allows the reader to see whether different nonparametric regression estimators produce similar or diverging fits to the same data. The example data come from Jacobson and Dimock's (1994) study of the 1992 US House elections. In the study of American politics, one well known regularity is that Congressional incumbents tend to be reelected. In the 1992 House elections, many incumbents were defeated, and Jacobson and Dimock explore the factors that contributed to the unusually high number of incumbents that were defeated that year. They argue that dissatisfaction with Congress was high due to a weak economy and a number of Congressional scandals. While they focus on a number of different predictors for the success of challengers, one indicator that they use to measure dissatisfaction with Congress is the percentage of the vote for H. Ross Perot in the 1992 presidential election. The district level vote between the president and members of the House is highly correlated, and Jacobson and Dimock test whether the level of support for Perot in this election increased support for challengers. For many voters, Perot served as a protest vote which may indicate dissatisfaction with incumbents of either party. In this illustration, we estimate a nonparametric regression model for the relationship between the challenger's vote share, as a percentage, and the percentage of the vote for H. Ross Perot in each Congressional district in the 1992 general election. Jacobson and Dimock assume the relationship between these variables is linear.[3] A nonparametric regression model based on local averages allows us to relax the linearity assumption and estimate the functional form from the data.

As a first step, it is useful to inspect a scatterplot between the two variables for obvious departures from linearity. Figure 2.1 contains a scatterplot of the two variables, and visual inspection of the scatterplot does not reveal an obvious nonlinear relationship between the two variables. A positive relationship between the two variables is, however, fairly obvious, and we could easily assume that

---

[3]It should be noted that they never consider the bivariate relationship between these two variables, but instead include it as one of several variables in a multivariate regression model.

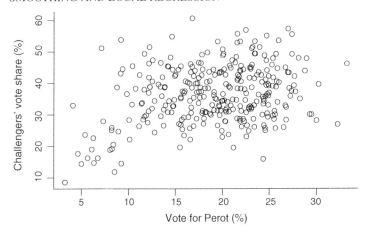

Figure 2.1   A scatterplot of challenger's vote share and Perot vote, 1992.

the relationship is linear. If we assume the functional form is linear and additive, it would be natural to regress the challenger's vote share on the support for Perot using least squares. The results of such an analysis reveal that the value of $\hat{\beta}$ is 0.46 with a standard error of 0.09, so the relationship is positive and highly significant. The parametric model assumes a globally linear relationship. Therefore, increases in the challenger's vote share should occur uniformly as support for Perot increases. This may not be a reasonable assumption. Perhaps, support for Perot has a diminishing effect on the challenger's vote share.

   If we use using a nonparametric estimator, we can relax the assumption of global linearity by using the moving average nonparametric estimator. Below is a comparison of the differing functional forms. For a linear regression model, the functional form is

$$\text{Challenger's vote share} = \alpha + \beta \text{ Perot vote} + \varepsilon. \tag{2.3}$$

For the nonparametric estimator, we write the functional form with more general notation

$$\text{Challenger's vote share} = f(\text{Perot vote}) + \varepsilon \tag{2.4}$$

where we assume that f is some smooth function. The moving average nonparametric regression model estimates the shape of f from the data without assuming a functional form. Of course, Equation (2.3) is a special case of Equation (2.4) when $\hat{f} = \alpha + \beta$. The fact that simpler parametric functional forms are nested within nonparametric estimates will prove to be very useful. In essence, we are trading the easily summarized estimate of β for a more flexible nonparametric fit.

The moving average nonparametric regression assumes that the values of x are in increasing order and has the form

$$\hat{f}(x_i) = \frac{1}{k} \sum_{j=\underline{i}}^{\bar{i}} y_j \tag{2.5}$$

where $\underline{i} = i - (k - 1)/2$ and $\bar{i} = i + (k - 1)/2$. In this notation, k represents the number of observations or width of bin. If k is set to 51, then 51 observations will be used to calculate the local average of y. By increasing or decreasing k, we increase or decrease the number of cases used to calculate the local average of y. The i notation refers to indices for the ordered values of x. For each average, x_i is the value of x at the focal point, while $x_{\underline{i}}$ and $x_{\bar{i}}$ represent the lower and upper values of x for each bin. Changing the value of i moves the bin along the range of x. Figure 2.2 contains a scatterplot of the two variables along with both the predictions from a linear fit and the moving average nonparametric regression estimate with bin width of 51 cases.

As is typical for moving average estimators, the fit is rather rough. The boundary bias is apparent as we are unable to calculate averages beyond the minimum and maximum values of the x variable. While the moving average fit does not appear to be a great departure from the linear fit, the effect of the Perot vote share does appear to reach a threshold between 10 and 15% and increases more slowly thereafter. This nonlinear feature is hidden by the global linear fit. Does changing the bin width have any effect on the nonparametric estimate? We can easily find out by estimating two new moving average estimates with different bin widths. Figure 2.3 contains three moving average estimates all

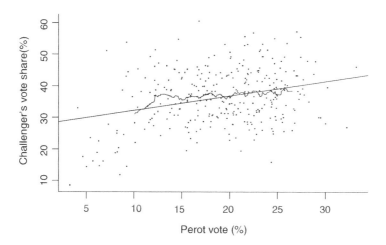

Figure 2.2   A moving average smoother and an OLS regression line.

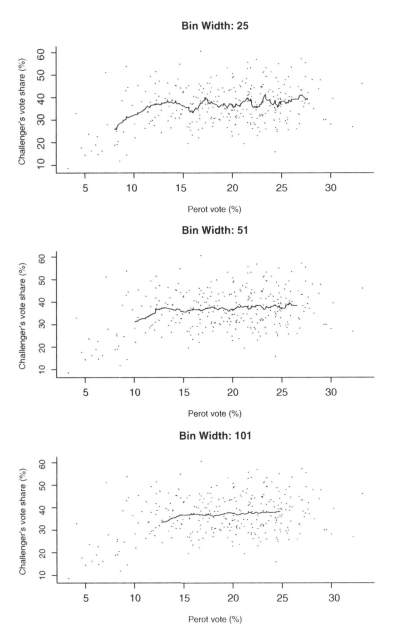

Figure 2.3   A moving average smoother: varying the bin width of the smoother.

using different bin widths. The estimate in the top panel uses a bin width of 25, and for this smaller bin width, there is less data in each bin, and the average we take at each point is less reliable so the estimate will exhibit greater variability. While the fit has a higher variance, it is more faithful to local variation in the data. The estimate in the middle panel uses a bin width of 51; identical to that used in Figure 2.2. In the bottom panel, the moving average estimate is based on a bin width of 101. As we expand the bin width, there is more data in each bin, and the fit is smoother, but it is less faithful to local variation in the data.

When the bin width is 25, the nonparametric estimate is highly variable. When it is 101, we see much less local variation. If the bin width is set to the sample size, the local estimates become global. The estimate with the smallest bin width has the most pronounced nonlinearity, while the estimate with the largest bin width is nearly flat and quite close to the linear regression estimate. For the estimate with the smallest bin width it appears that the effect increases until a threshold around 15% is met; after this threshold, the effect is not as strong. Notice also that increasing the bin width increases the boundary bias, as we have less data at the margins. In this example, here, the boundary bias appears to make the nonparametric estimate appear closer to a linear fit. While the moving average nonparametric estimator is useful for demonstrating the basic principles of nonparametric regression, it lacks one further refinement required for better estimates.

2.1.2 Kernel Smoothing

Thus far, for the local estimates, we have used averages of y. Simple averages are, however, unsuitable for local estimates. The goal with nonparametric regression is to produce an estimate that is faithful to the local relationship between x and y. With simple averages, we give equal weight to values of x and y no matter their distance from the focal point. The use of an unweighted average is particularly problematic with the moving window bin, as observations are used repeatedly in the local averages. It makes little sense to give all observations the same weight regardless of the distance between an observation and the focal point. If we weight observations near the focal point more heavily than observations near the boundary of the bin, we can produce a nonparametric estimate that is more faithful to the local relationship between x and y.

Kernel smoothing is a nonparametric regression model that refines moving average smoothing through the use of a weighted average. What we need is a function that will weight data close to the focal point more heavily that observations farther from the focal point. Before considering the form of the weighted average, we require a measure of distance from the focal point. We will use this

measure of distance to assign weights in the weighted average. A useful measure
of distance is

$$z_i = \frac{(x_i - x_0)}{h}.$$

(2.6)

The term z_i measures the scaled and signed distance between the x-value for the
ith observation and the focal point: x_0. The scale factor, h, is called the bandwidth
of the kernel estimator, and it controls the bin width. Of course, this implies that h
controls how smooth or rough the nonparametric estimate will be. As one might
imagine, smaller values of h will produce rougher fits; very small values of h
will cause the nonparametric estimate to interpolate between each value of y. On
the other hand, larger values of h will produce smoother fits that may miss some
local detail.

We now need a function to assign the actual weights to the data. We will refer
to the weighting function as a kernel function: $K(z)$. The kernel function attaches
the greatest weight to the observations that are close to x_0, observations that have
a small value for z_i, and the function applies lesser weights symmetrically and
smoothly as the value of $|z|$ grows. The kernel function is applied to z_i to calculate
a weight for each observation that is in turn used to calculate a local weighted
average. A set of weights for a single bin result from the kernel function K being
applied to the signed and scaled distances within each bin:

$$w_i = K[(x_i - x_0)/h].$$

(2.7)

The weights, w_i, are then used to calculate to the local weighted average:

$$\hat{f}(x_0) = \frac{\sum_{i=1}^{n} w_i y_i}{\sum_{i=1}^{n} w_i}.$$

(2.8)

Thus far, we have not defined the kernel function. In principle, there are a
number of different kernel functions that we might use to calculate the weights
used in the local averages. Whatever the weighting function, the weights must
have certain properties. First, the weights must be symmetric about the focal
point; second, the weights must be positive, and we desire the weights to decrease
smoothly from the focal point to the bin boundary. Several different functions
meet these criteria, but the most commonly used kernel is the tricube kernel:

$$K_T(z) = \begin{cases} (1 - |z|^3)^3 & \text{for } |z| < 1 \\ 0 & \text{for } |z| \geq 1. \end{cases}$$

(2.9)

One advantage to the tricube kernel is that it is easily calculated. This was one of
its chief recommendations for early smoothing applications. The ease of compu-
tation matters little now, but this kernel is still widely used. Another commonly

used kernel is the Gaussian or Normal kernel, which is simply the Normal density function applied to the values of z_i:

$$K_N(z) = \frac{1}{\sqrt{2\phi}} e^{-z^2/2}. \tag{2.10}$$

The difference between nonparametric fits with the tricube and Gaussian kernel is typically neglible. It is instructive to note that the simple moving average smoother is also a kernel smoother. The moving average smoother has a rectangular kernel. For the rectangular kernel, we give each observation in the window equal weight which produces an unweighted local average:

$$K_R(z) = \begin{cases} 1 & \text{for } |z| < 1 \\ 0 & \text{for } |z| \geq 1. \end{cases} \tag{2.11}$$

In the illustration of kernel smoothing we return to the Congressional election example. Thus far, we have only produced crude nonparametric estimates of the relationship between the challengers' vote share and the level of support for H. Ross Perot. We can use a kernel smoother to produce more refined nonparametric regression estimates. Before estimating the kernel smoother estimate, we must first choose a value for h, the bandwidth. The method for choosing h is typically one of visual trial and error, where the analyst chooses the smallest value of h that produces a smooth fit. The scale for h depends on the software used for kernel estimation, but h is usually expressed on a numeric scale where larger values increase the bandwidth. Again, a larger bandwidth will produce smoother fits, while smaller bandwidths will produce rougher fits.

Within each bin, we calculate the scaled distance of each x from the focal point, apply the kernel function to calculate weights, and use these weights to estimate the local weighted average. Some graphics are useful to illustrate the operation of kernel smoothing. Figure 2.4 contains an example bin for the Congressional elections data. The solid line represents the focal point and the dashed lines represent the edges of the bin. For the moving average estimator, we would take the average of the y values within this bin, but for a kernel smoother, we first weight observations close to the focal point.

Finally, Figure 2.5 contains a plot of the tricube weights within the example bin from Figure 2.4. The observation at the focal point receives the greatest weight, a one, while observations near the edge of the bin are given very little weight. By moving the bin across the range of x, applying the tricube function and calculating the local weighted average, we produce a nonparametric estimate that is more faithful to local variation and smoother than a simple moving average estimate.

Figure 2.6 contains three different kernel smooth estimates of the relationship between support for challengers and support for Ross Perot using the tricube

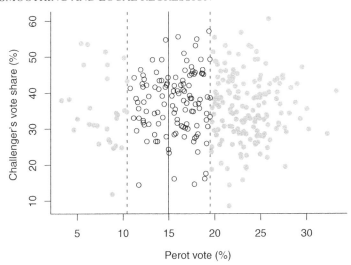

Figure 2.4   An example bin for the Congressional vote data.

kernel. For each estimate, we used bandwidth values of 4, 8, and 12. We should expect smoother fits for larger bandwidths. The kernel estimate more readily reveals what appears to be a nonlinear relationship between support for Perot and support for challengers. The relationship is strong for lower levels of support

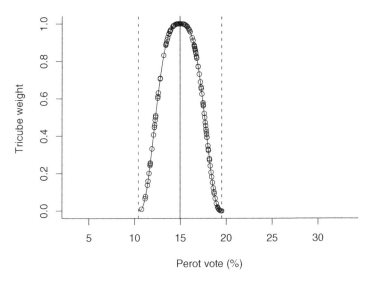

Figure 2.5   The tricube weighted average.

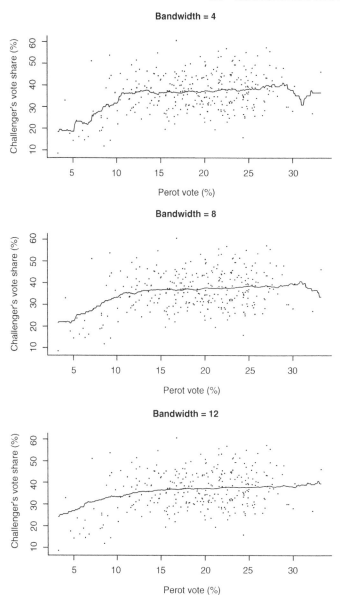

Figure 2.6   Kernel smoother with three different bandwidths.

for Perot, but the dependency levels off once support for Perot reaches around 15%. As is readily apparent, changing the bandwidth alters the nonparametric fit. The smaller bandwidths tend to give greater emphasis to the nonlinearity in the relationship. But this greater faithfulness to the local relationship comes at a cost of added variability that is missing when the bandwidth is set to 12. Regardless of which bandwidth is used, the nonlinearity revealed should cause an analyst to question whether a global linear fit between the two variables is an adequate model of the relationship.

While kernel smoothers are an improvement over simple moving average estimators, they still have serious drawbacks. As we will see, the main drawback to the kernel smoothers is that the mean, weighted or otherwise, is not an optimal local estimator. As such, kernel smoothers are best for pedagogical purposes as opposed to being used in applied work. While the concepts of weights and moving bins will remain important, use of a local regression estimate instead of a local mean produces a superior nonparametric regression model.

2.2   Local Polynomial Regression

Thus far we have calculated the local estimate of $y|x$ using the mean as the estimator. One might ask whether there is any reason to use the mean rather than some other estimator. The main reason to use the mean as the local estimator is for ease of computation. Nonparametric regression was long considered to be a computationally intensive statistical technique since it requires estimation of many local estimates. Before computing was as powerful as it is presently, the ease of estimating the mean repeatedly was important. Modern computing, however, has rendered such concerns moot. In fact, there are reasons to prefer other local models. While using the mean for local estimation is useful for pedagogical purposes, we can do better in terms of statistical performance with other local estimators.

One obvious alternative to the mean as a local estimator is least squares. In fact many widely used nonparametric regression models are based on local *regression* estimates instead of local mean estimates, since regression has bias reduction properties that the mean does not. The basic algorithm for the nonparametric regression estimate remains nearly identical to that for kernel smoothing. Again, we construct a series of bins with widths determined by the analyst. Within each bin, we regress y on x, and the local estimate is the predicted value for y at the focal point x_0. This process is repeated for each bin. Joining the predicted values from each bin with line segments produces the nonparametric estimate of f. A visual demonstration of local regression is useful at this point.

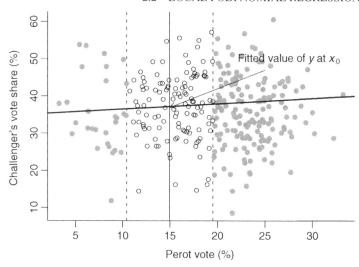

Figure 2.7   Local linear regression: fitted Y at the focal point.

Figure 2.7 contains a scatterplot of the Jacobson data with markers for the bin and a local regression line. In this plot the window width remains the same as for Figure 2.4, but instead of estimating a mean for the data within this bin, we estimate a regression between the two variables with only the data points that fall within the bin, represented by the circles in Figure 2.7. The fitted value of y calculated at the focal point, or more informally the point where the regression line crosses the value of the focal point becomes the value from the local regression that contributes to the nonparametric estimate. The window would then shift one observation to a new focal point, and we would refit the local regression with the data from this new bin.

Cleveland (1979) first proposed the local regression smoother and most statistical software uses the same basic algorithm he proposed. His local regression methods *loess* and the variant *lowess* are among the more widely used nonparametric regression models. Both of these smoothers complicate the idea of local regression outlined thus far in several ways. Given the ubiquity of these smoothing methods, we now turn to a more detailed discussion of Cleveland's nonparametric regression models.

Cleveland proposed a polynomial regression model instead of a linear fit for the local estimate. For polynomial regression, a p-degree polynomial in a predictor x is

$$y = \alpha + \beta_1 x + \beta_2 x^2 + \ldots + \beta_p x^p + \varepsilon. \tag{2.12}$$

If $p = 1$ then the fit is linear, if $p = 2$ then the fit is quadratic, and so on. The choice of polynomial degree is left to the analyst, though polynomials of an order higher than two typically do little to enhance the nonparametric estimate. The following iterative fitting process produces the nonparametric regression estimate. First, a set of weights are calculated for each x observation within the bin. We calculate these weights identically to those used for kernel smoothing. That is, we calculate the signed and scaled distance of each observation from the focal point, and a kernel function is applied to these distances to produce a set of symmetric weights for each observation with the greatest weight given to the focal point. Cleveland recommended the use of the tricube kernel. Next, the local polynomial regression model is fit within the bin using weighted least squares and the weights, w_i. Specifically, we estimate the following local regression model:

$$\frac{y_i}{w_i} = \alpha + \beta_1 \frac{x_i}{w_i} + \beta_2 \frac{x_i}{w_i}^2 + \ldots + \beta_p \frac{x_i}{w_i}^p + \frac{\varepsilon_i}{w_i} \qquad (2.13)$$

which minimizes the weighted residual sum of squares: $\sum_{i=1}^{n} w_i e_i^2$. While the tricube kernel is typically used, other weights are possible. After the initial weighted least squares fit, a new set of robustness weights based on the residuals from the local fit are calculated. Observations with large residuals are given small weights, and observations with small residuals are given large weights. These robustness weights are:

$$\delta_k = B(e_k/6s) \qquad (2.14)$$

where e_k is the local residuals, s is the median of $|e_k|$, and B is a bisquare kernel function, which is equivalent to the tricube kernel except a second order polynomial is used. A new local fit is estimated repeatedly with an updated set of weights until the change in local estimates is within a defined level of tolerance. Finally, the fitted value for y is calculated at x_0. The process is then repeated within each bin, and the set of fitted values are joined with line segments to form the nonparametric estimate. The local regression procedure can be done without weights. In fact, the absence of weights defines the difference between *loess* and *lowess*. The first is short for *local regression*, while the latter is short for *local weighted regression* Cleveland (1993). The differences between the two is typically minimal in practice.

Selection of the bandwidth remains important for local polynomial regression (LPR), since the bandwidth, as before, controls how smooth the overall fit is. Again, a wider bandwidth produces smoother fits between x and y, and a smaller bandwidth produces a rougher fit that if small enough interpolates the relationship between the two variables. If a single bin is used, the LPR estimate will be identical to a global least squares fit. One notational difference in bandwidth selection is common. Instead of specifying the bandwidth in terms of the number of observations on either side of the focal point, it is more convenient to specify the bandwidth as the proportion of the observations that will be included in the window. This fraction of the data is called the span, s, of a local regression smoother. The number of observations included in each window is then $m = [sn]$, where the square brackets denote rounding to the nearest whole number and n is the total number of observations. For example, there are 312 cases in the Jacobson data. If we set the span, s, to 0.50 (a typical value), there will be 156 (0.50×312) data points within each bin.

Why should we prefer the local regression model to local averaging? Fan and Gijbels (1992) and Fan (1993) demonstrated that the local linear fit has several excellent properties that local averaging does not. They prove that local polynomial regression fits reduce bias in $\hat{f}$ compared to nonparametric regression based on local averaging. The bias reduction properties of local polynomial regression (LPR) is strongest at the boundaries of the relationship between x and y, where the kernel estimator tends to flatten the nonparametric fit.

Local Polynomial Regression: An Example

We now return to our example from the 1992 House elections. Thus far, we have seen moving average and kernel nonparametric estimates of the relationship between the challenger's vote share and support for Perot. The nonparametric estimate from a kernel smoother revealed a threshold effect between the two variables where support for Perot had a strong effect on challengers' vote share before leveling off. Figure 2.8 contains plots of both *loess* and *lowess* estimates with the span set to 0.5, which means that 50% of the data are used in each bin. The plot also contains a global linear fit estimated with least squares. We see very little difference between the two nonparametric fits as both reveal the same nonlinear relationship between challenger's vote share and support for Perot.

How do we interpret the results? With a standard linear regression, we would focus on $\hat{\beta}$, but with nonparametric regression no single parameter estimates summarize the relationship between x and y. For LPR, the plot summarizes the relationship between the two variables. Visual inspection of the plot clearly demonstrates that the challenger's vote share increases steeply with the level of support for Perot until a threshold of approximately 15% is reached. At this point, the relationship levels off and is flat. The nonparametric estimate reveals a common form of nonlinearity: the threshold effect. For a threshold effect, a strong relationship between two variables reaches a threshold and either levels off or begins to decline. Contrast this with the global linear fit. Recall that for a linear

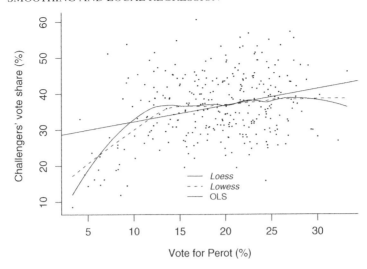

Figure 2.8 *Loess* and *lowess* fit to challenger's vote share and support for Perot, 1992.

regression model fit between these two variables, the estimated β was 0.46. The linear model overpredicts the vote share for challengers in districts where support for Perot was the strongest. The linear model would predict that the challenger would receive 43% of the vote in the district with the strongest support for Perot. Under the nonparametric model, we would predict the challenger's vote share to be 38% in the districts with the strongest support for Perot. The linear fit substantially overpredicts the challenger's votes share in districts where Perot enjoyed little support. For example, in a district where Perot received 5% of the vote, the global fit would predict the challenger's vote share to be approximately 30%. The *loess* prediction of the vote share for the challenger for the same level of Perot support is 19%. While the global linear fit provides easy interpretation, the functional form for this model appears to be incorrect which leads to a misleading picture of the relationship between challengers' success and support for Perot.

The assumptions that underlie the nonparametric estimate are less restrictive than those for OLS as there is no strong assumption of global linearity. When estimating a global linear fit, the analyst must assume a functional form, but nonparametric regression estimates the functional form from the data. In this illustration, the nonparametric estimate should make any analyst pause before estimating a linear regression model. Given the differences between the nonparametric regression models and the global linear model, we have some evidence that the assumption of a linear functional form is incorrect.

Local Likelihood

We can further generalize the concept of local regression. Loader (1999) generalizes local estimation with local likelihood, which allows the analyst to estimate a number of different local models. For example, using local likelihood, one can estimate local generalized linear models such as logistic or poisson regression. In practice, there is usually little need for such methods, since we can add smoothing to GLMs in the scale of the linear predictor. Interested readers should consult Loader (1999).

2.3   Nonparametric Modeling Choices

Nonparametric regression allows the analyst to make minimal assumptions about the functional form of the statistical relationship between x and y, since the functional form is estimated from the data. This is a marked departure from the familiar linear regression model which requires the analyst to assume *a priori* a linear functional form. While we have relaxed the functional form assumption, this comes at some cost since the analyst must make several modeling choices not required for parametric models. At a minimum, the analyst must choose both the degree of polynomial for the local regression models and a value for the span. If using *lowess*, one can also vary the weight functions for the local estimate. In this section, we consider how these choices affect the nonparametric regression estimate. We would prefer that the nonparametric estimate be invariant to these choices, and, typically, nonparametric regression is robust to the choice of model settings. The reader will see that the choice of the span is the most important modeling choice. We have already seen that the use of local weights made little difference in the Congressional elections example. The unweighted *loess* fit differed little from the weighted *lowess* estimate. The order of the polynomial for the local regression estimate also tends to matter little in practice. We start with how the choice of span affects the nonparametric regression estimate.

2.3.1   The Span

The choice of the span is the most critical choice an analyst must make when using nonparametric regression. Why is this? For many statistical models, we often trade a reduction in bias for increased variance, or we trade a reduction in variance for increased bias. With nonparametric regression smoothers, this tradeoff is encapsulated by the choice of span. If the span is too large, the nonparametric regression estimate will be biased, but if the span is too small, the estimate will be overfitted with inflated variance, which will affect the estimated confidence bands. At the outset, it might appear that one must exercise great care in choosing the span or the nonparametric estimate will be problematic. While some care is required, the overall nonparametric regression estimate is usually invariant to a wide range of span values, and bias and overfitting are only possible for extreme values of the span.

An illustration with simulated data allows for a demonstration of how choosing the span embodies a bias–variance tradeoff. The simulated data generating process in this illustration is one where y is a highly nonlinear function of x.[4] We use *lowess* to estimate the nonlinear relationship between the simulated x and y.[5] We estimate three different *lowess* fits with span settings of 0.0001, 0.3, and 0.6; for the first fit, we use less than 1% of the data, and for the second and third fits, we use 30% and 60% of the data in each bin. In this example, we include 95% confidence bands. While we have yet to cover inference for these models, the affect on the variance is most apparent in the confidence bands. We take up the topic of inference shortly. Figure 2.9 contains the plots of the three *lowess* estimates and confidence bands, and the solid vertical lines represent the bin width from each span setting.

In each panel, the dashed line represents the true regression function. The top panel of Figure 2.9 contains a plot of the *lowess* fit with a span setting of 0.0001. With such a small span setting, the local estimate is very faithful to the local relationship between x and y, but since there is little data within each bin, the estimate also displays considerable variability. In particular, note that the confidence bands are quite wide. The result is a nonparametric regression estimate with little bias but inflated variance and wide confidence bands. Here, we have overfit the data using more local parameters than is necessary to summarize the fit between x and y. In the context of nonparametric regression, we call such overfitting undersmoothing. The second panel of Figure 2.9 contains a plot of the *lowess* fit with a span setting of 0.30. The *lowess* fit, here, displays much less variability, but at the peaks and valleys does not quite match the true regression function. We have decreased the variability from the first fit, but the estimate is now slightly biased. We call these departures from the true regression function oversmoothing. The reader should note, however, that the bias does not change the basic estimate of the nonlinear functional form. We would not confuse the estimate for some other nonlinear form unrepresentative of the true regression function. The third panel of Figure 2.9 contains a plot where the span setting is 0.60, which implies that 60% of the data is used for each local estimate. The

---

[4]The exact function is $y = x \cos(4\pi x) + 0.3\epsilon$, where x is a sequence from 1 to 50 and $\epsilon \sim N(0, 1)$.

[5]Why *lowess* instead of *loess*? Mainly for reasons of convenience.

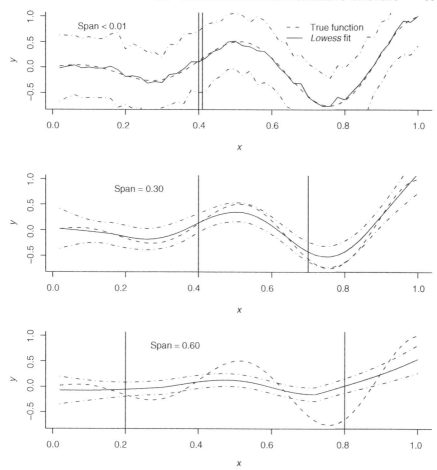

Figure 2.9   The bias variance tradeoff in smoothing.

lowess estimate now exhibits severe oversmoothing. Now, however, there is very little variance given that we are using most of the data for each local estimate. The price for reducing the variance is that the *lowess* fit is not very faithful to the true regression function. Here, the bias obscures several notable features in the true functional form. Obviously, if we used 100% of the data for each local fit, the fit would cease to be local as the *lowess* fit would be identical to a global linear fit to the data.

In selecting the span, we must choose between oversmoothing which induces bias and undersmoothing which is more precise but can overfit the data and produces a nonparametric fit that is too variable. Oversmoothing can distort the true regression function and undersmoothing affects the estimation of confidence bands. The goal in span selection is to choose the span that produces as smooth as fit as possible without distorting the underlying pattern in the data (Cleveland 1993). The ideal span parameter minimizes *both* the variance and the bias, which suggests a mean squared error criterion for span selection. While there are methods for estimating the optimal span setting, these methods introduce a number of complications which are explored in Chapter 4. The easiest and most commonly used method for span selection is visual trial and error. Standard practice for the trial and error method is to start with a span setting of 0.50. If the fitted nonparametric regression estimate looks too rough, we increase the span by a small increment with an increase of 0.10 being a reasonable increment. If the fit still appears too rough, we increase the span again. If either the initial span setting or an increase in the span produces a smooth looking fit, one should see if the span can be decreased without making the fit overly rough.

To illustrate span selection with real data, we return to the example from the 1992 House elections. Figure 2.10 contains three plots of *lowess* fits with different span settings. For the first *lowess* estimate, in the top panel, we start with the recommended 0.50 span setting. The resulting nonparametric estimate is smooth and does not appear to be overfit. The question is: can we decrease the span and retain the same level of smoothness? For the next fit, we decreased the span to 0.40, and the result is in the second panel of Figure 2.10. Here, the overall fit changes little, so we reduce the span by a further 0.10 increment to 0.30. The result, in the third panel of Figure 2.10, displays some very slight increased variation. Further decreasing the span to 0.20 produces an obviously overfit estimate. Importantly, the *lowess* estimates are fairly robust to the choice of span. Regardless of the span value chosen, we still see the same threshold effect where it is clear that support for Perot has little effect on the challenger's vote share once a threshold is reached. Often it takes fairly extreme span values to either overfit the data or to oversmooth the estimate. The visual span selection method is a bit *ad hoc* but works well in practice. In general, the analyst should err on the side of undersmoothing rather than oversmoothing. We use nonparametric regression since the global linear fit oversmooths the relationship between x and y, and it is usually obvious when the estimate is overfit but less obvious when it is oversmoothed.

2.3.2 Polynomial Degree and Weight Function

Besides the span, the analyst can use different polynomial degrees for the local regression model, and, if using *lowess*, he or she can select differing initial weight

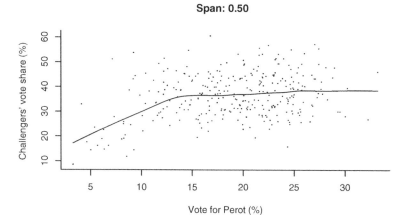

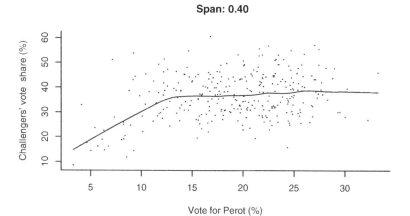

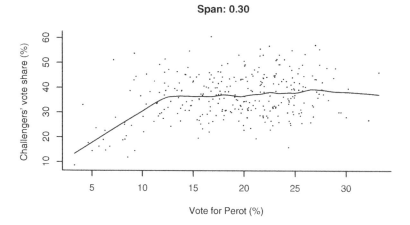

Figure 2.10   Selecting the span for a *lowess* fit to the data.

functions. These two choices have a far smaller effect on the nonparametric fit than that of the span. In fact, typically there is little need to adjust either of these settings. First, we examine the effect of changing the polynomial degree.

A higher degree polynomial will provide a better approximation of the underlying mean than a lower degree polynomial, that is a higher degree polynomial will have less bias. But a higher degree polynomial model has more coefficients resulting in greater variability and will tend to overfit the data. In fact, the polynomial degree and span have confounding effects on the nonparametric regression estimate. If we hold the span constant, and compare local linear fits to local quadratic fits, the quadratic fit will be more variable, but one can compensate for the increased variability by adjusting the span. Standard practice is to choose a low degree polynomial and use the span to adjust the overall fit (Loader 1999). In fact, Cleveland (1993) does not entertain local polynomial regressions of an order higher than two, since local fits of a higher order rarely improves the fit but uses extraneous parameters. Local linear fits have difficulty if there are multiple local maxima or minima in the data, but such forms of nonlinearity are rare in the social sciences. In actual practice, the degree is usually set to quadratic and not adjusted. In fact, many implementations of *lowess* and *loess* do not allow the analyst to adjust the polynomial degree above quadratic if at all.

Figure 2.11 contains plots of the *lowess* estimate for challenger's vote share and support for Perot where we hold the span constant at 0.50, but use local linear, quadratic, and cubic fits. The overall estimates are quite similar, but we can see some slight variation across the three estimates. The local linear fit displays slightly less curvature around the threshold than either the local quadratic or cubic estimates, but the differences between the quadratic and cubic fits are minimal. As a practical matter, one might experiment with linear and quadratic fits if the software allows such an adjustment. In general, there will be little difference between a linear and quadratic fit with most social science data. If there is little difference between the two, the analyst should err on the side of parsimony and use a local linear fit.

Finally, one can choose different weight functions for *lowess* smoothers. The weight function has little effect on the bias–variance tradeoff, but it can affect visual quality. The most popular weight is the tricube weight, and it is the default for most software programs. For most applications, there is no reason to experiment with different weights.

2.3.3   A Note on Interpretation

At this point, it is useful to make some general remarks about the interpretation of nonparametric plots. When examining nonparametric estimates, it is invariably

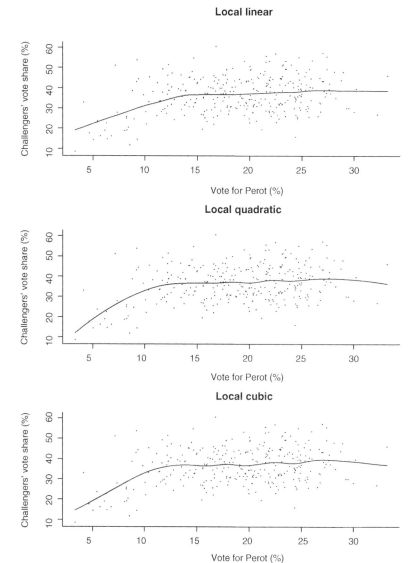

Figure 2.11   Varying the polynomial degree for local polynomial regression.

tempting to try and impose some meaning on small bumps or curves in the fit. While this might be tempting, it is a temptation analysts should resist. Such undulations are often the result of some highly local relationship between x and y. Such bumps can also be a function of overfitting the relationship between the two variables. Overfitting means that the fitted nonparametric function is following small and probably random features in the data. Figure 2.12 contains a *lowess* estimate of the relationship between challenger's vote share and support for Perot where we have set the span to 0.20. In this estimate, obvious overfitting occurs as we see considerable variation in the estimate once support for Perot is higher than 15%. In nonparametric regression estimates such as this, there is a surplus of fit. When there is a surplus of fit, the overall pattern in the data remains, but we see local variation that is not supported by most social science theory. As such, we would not expect these highly local forms of nonlinearity to be present in out of sample contexts.

Overfitting is a common criticism of nonparametric regression estimates. While it is true that one can overfit the data with nonparmetric regression, overfitting is possible with any statistical model. Moreover, there are several steps one can take to avoid overfitting. First, analysts should examine the overall pattern of the nonlinearity and not focus on small features in the nonparametric estimate.

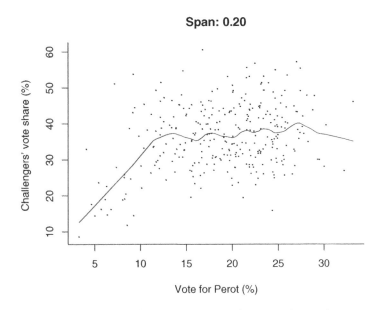

Figure 2.12   Overfitting a nonparametric regression estimate.

It is useful to ask questions like: 'Is there some clear threshold where the relationship between x and y changes?' Here, it is also useful to compare different
types of smoothers to ensure that the same general form of nonlinearity appears
regardless of how one estimates the nonparametric fit. An analyst can also err
on the side of oversmoothing. By increasing the span some small amount, local
variation is typically smoothed out which prevents overfitting. Recall that for this
example an increase of 0.10 in the span removes all the local variation in the
estimate. That said, unless we see highly local variation as in Figure 2.12, one
should err on the side of undersmoothing to avoid a biased estimate.

2.4 Statistical Inference for Local Polynomial Regression

With parametric models, statistical inference focuses on estimating standard
errors for the model parameters. The analyst uses the standard errors in conjunction with a distributional assumption to form confidence intervals and perform
hypothesis tests. Despite the fact that nonparametric regression does not produce
a single parameter that summarizes the statistical relationship between x and y,
statistical inference proceeds in a similar fashion. With nonparametric regression
models, we can estimate confidence bands for the nonparametric estimate, perform hypothesis tests, and test the nonparametric regression model against other
nested models such as the global linear model.

Confidence Bands

The process for estimating confidence bands is similar to the estimation of confidence intervals for predictions from linear regression models. That this is the
case should not be surprising since the LPR estimate is comprised of the fitted
values from a series of local regressions. We can express each fitted value from
the nonparametric regression model as a linear combination of the y_i values:

$$\hat{y}_i = \sum_{k=0}^{d} \beta_k x_i = \sum_{k=1}^{n} s_{ij}(x_i) y_i. \tag{2.15}$$

The fitted value $\hat{y}_i$ results from a locally weighted least squares regression of y_i on
a subset of x values. The term $s_{ij}(x_i)$ denotes the coefficients from each the local
regression as the weights that along with the y_i values form the linear combination.
If we assume that the y_i's are independently distributed with a constant variance,
then the following is true: $V(y_i) = \sigma^2$. Under this assumption of constant variance,
the variance of the fitted values which comprise the nonparametric regression

estimate is

$$V(\hat{y}_i) = \sigma^2 \sum_{j=1}^{n} s_{ij}^2(x_i). \tag{2.16}$$

At this point, it is more convenient to express the variance of $\hat{y}_i$ in matrix notation. First, we rewrite the equation for the predicted values of y in matrix form

$$\hat{\mathbf{y}} = \mathbf{S}\mathbf{y} \tag{2.17}$$

where $\mathbf{S}$ is a $(n \times n)$ smoothing matrix whose (i, j) elements are $s_{ij}^2(x_i)$, the local regression coefficients at each focal point. The variance for the fitted values is now easy to derive as it takes a form very similar to that of a least squares linear regression:

$$V(\hat{\mathbf{y}}) = \mathbf{S}V(\mathbf{y})\mathbf{S}' = \sigma^2 \mathbf{S}\mathbf{S}'. \tag{2.18}$$

The reader will perhaps have noticed the similarity between $\sigma^2\mathbf{S}\mathbf{S}'$ and $\sigma^2\mathbf{X}\mathbf{X}'$ the variance of y from a linear regression model.

Of course, we still require an estimate for σ^2. We can further exploit the similarity between nonparametric regression and linear regression to estimate σ^2. With OLS, the squared and summed residuals provide an unbiased estimator of σ^2:

$$\hat{\sigma}^2 = \frac{\sum e_i^2}{n-2} \tag{2.19}$$

where $e_i = y_i - \hat{y}_i$ is the residual for observation i, and $n - 2$ is the degrees of freedom associated with the residual sum of squares. We use a similar estimator for σ^2 in the nonparametric setting. The numerator is nearly identical as the residuals are easily calculated from $y_i - \hat{y}_i$, where $\hat{y}_i$ is the predicted value for from each local regression, but to complete the analogy, we require an equivalent calculation for the degrees of freedom for the residual sum of squares. We again rely on an analogy between linear regression and nonparametric regression to calculate the residual degrees of freedom. In the context of linear regresion, the number of parameters in the model can be calculated as the trace of the hat matrix $\mathbf{H}$ where $\mathbf{H} = (\mathbf{X}'\mathbf{X})^{-1}\mathbf{X}'$. The $\mathbf{H}$ matrix is often referred to as the 'hat' matrix since it transforms the observed y values to a set of predicted values: $\hat{\mathbf{y}} = \mathbf{H}\mathbf{y}$. The trace of $\mathbf{H}$ is equal to $k + 1$, the number of the parameters in the model, and we calculate the residual degrees of freedom for the linear model as $\mathrm{tr}(\mathbf{I}_n) - \mathrm{tr}(\mathbf{H}) = n - (k + 1)$, where $\mathbf{I}_n$ is an $n \times n$ identity matrix.

In the context of nonparametric regression, the $\mathbf{S}$ matrix is equivalent to $\mathbf{H}$, since the local regression parameters transform the y_i values to fitted values. We use the $\mathbf{S}$ matrix to calculate both the degrees of freedom and the residual degrees of freedom for a nonparametric regression model. For a nonparametric regression model, $\text{tr}(\mathbf{S})$ is the effective number of parameters used by the smoother. Unlike in a linear model, however, the degrees of freedom need not be integers. The residual degrees of freedom, df_{res}, are: $\text{tr}(\mathbf{I}_n) - \text{tr}(\mathbf{S}) = n - \text{tr}(\mathbf{S})$.[6] Therefore, the estimated error variance is

$$\hat{\sigma}^2 = \frac{\sum e_i^2}{n - \text{tr}(\mathbf{S}),} \tag{2.20}$$

and the estimated variance of the fitted values is

$$V(\hat{\mathbf{y}}) = \hat{\sigma}^2 \mathbf{S}\mathbf{S}'. \tag{2.21}$$

Along with a distributional assumption, we can now form pointwise confidence bands for the nonparametric regression fit. If we assume normally distributed errors, or if we have a sufficiently large sample, a pointwise approximate 95% confidence interval for $\hat{y}$ is $\pm 2\sqrt{s_{ij}}$, where s_{ij} corresponds to the the diagonal elements of $\mathbf{S}\mathbf{S}'$ when $i = j$. The 95% confidence bands for each $\hat{y}_i$ value can be plotted along with the nonparametric estimate. When the bands are close to the line, we have a highly precise fit; if the bands are far apart the fit is not very precise. The width of the confidence bands is, of course, highly dependent on the amount of data around each focal point.

Hypothesis Tests

We can also conduct several different hypothesis tests for nonparametric regression models, since the nonparametric estimate is a general model that nests more parsimonious parametric models. As such, F-tests provide a natural framework for conducting hypothesis tests with these models. The standard hypothesis test conducted with a linear regression model is whether the population slope, β, is equal to zero ($H_0 : \beta = 0$). By convention, for p-values less than 0.05, we reject this null hypothesis. We can conduct a similar hypothesis test with the nonparametric regression model using an F-test. The F-test of no relationship in the nonparametric context takes the following form:

$$F = \frac{(\text{TSS} - \text{RSS})/(df_{\text{mod}} - 1)}{\text{RSS}/df_{\text{res}}} \tag{2.22}$$

where $\text{TSS} = \sum(y_i - \bar{y}_i)^2$ is the total sum of squares. The term $\text{RSS} = \sum(y_i - \hat{y}_i)^2$ is the residual sum of squares from the fitted nonparametric model, $df_{\text{mod}} =$

---

[6]Hastie and Tibshirani (1990) provide details on alternatives to using $\text{tr}(\mathbf{S})$ to calculate the degrees of freedom. See Chapter 3 of Hastie and Tibshirani (1990) for details. These alternatives are typically more difficult to compute and see little use.

$\text{tr}(\mathbf{S}) - 1$, and $df_{\text{res}} = n - \text{tr}(\mathbf{S})$. In short, we test the nonparametric model against a model with only a constant. The resulting test statistic can be used to calculate a p-value, which if it is less then the standard 0.05 threshold, we reject the null hypothesis. If we are unable to reject the null hypothesis, it is typically assumed that there is no relationship between x and y. Such hypothesis tests are in keeping with standard practices of data analysis in the social sciences. However, visual examination of the nonparametric estimate is often sufficient to determine if the two variables are unrelated.

While standard hypothesis tests are of interest to many researchers, a more important test is one where we test the nonparametric model against a parametric alternative. For example, we can test the nonparametric fit against the more parsimonious linear regression model. If we test a nonparametric regression model against a linear model and find they fit the data equally well, we can conclude that a linear functional form adequately describes the relationship between x and y. If the nonparametric model provides a better fit to the data, we conclude that the linear functional form is incorrect, and we need to model the nonlinearity between x and y. We interpret such a test as a test for a statistically 'significant' amount of nonlinearity in the relationship between x and y. Importantly, we can also test the nonparametric fit against power transformations or different nonparametric regressions with differing numbers of parameters. The nonlinearity test is based on an approximate F-test since the linear model is a special case of the more general nonparametric regression model. Denoting the residual sum of squares from a more restrictive model (linear, quadratic, etc.) as RSS_0 and the residual sum of squares from the nonparametric regression model as RSS_1, the test statistic is

$$F = \frac{(\text{RSS}_0 - \text{RSS}_1)/J}{\text{RSS}_1/df_{\text{res}}} \qquad (2.23)$$

where J is the difference in the number of parameters across the two models. When comparing a linear fit to a nonparametric fit, J is $df_{\text{mod}} - 2$. The resulting test statistic is F-distributed with the degrees of freedom equal to df_{mod} and df_{res}, and if the calculated p-value is less than 0.05, the analyst prefers the nonparametric regression model. This test allows the analyst to easily test between nonparametric models and models with global fits. While models with global assumptions such as those with linear functional forms or power transformations might be acceptable, we often have little evidence to justify the functional form assumption. The nonparametric regression model supplies the analyst with a convenient means of testing such assumptions. The nonparametric regression model is often not a replacement for parametric models, but it does provide a means for testing parametric assumptions.

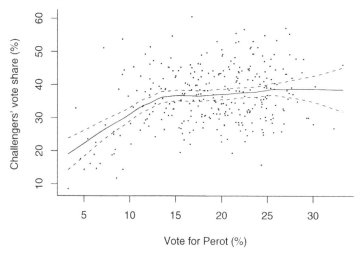

Figure 2.13   95% confidence bands for the *loess* fit to the challenger's vote share data.

Statistical Inference: An Example

Thus far, we have seen several nonparametric regression estimates of the relationship between support for Perot and the challenger's vote share in the 1992 House elections. The nonparametric regression estimates revealed a nonlinear dependence between these two variables. Using the tools of statistical inference, we can make several probabilistic statements about the nonparametric estimate. First, we estimate pointwise 95% confidence bands. Figure 2.13 contains a locally linear *loess* estimate with the span set to 0.50 along with pointwise confidence bands.[7] As is often the case, the confidence bands widen at the boundaries of the data. A rough visual test between the nonlinear and linear fit is to see if one can draw a straight line within the confidence bands.

Of course, we can also perform hypothesis tests. First, we test whether support for Perot has a statistically significant effect on challenger's vote share. Here, we use an F-test to compare the nonparametric model against modeling challenger's vote share as a function of a constant. The total sum of squares is 25 684.06, and the residual sum of squares from the *loess* fit is 21 811.94. The model degrees of freedom are 5.14, and the residual degrees of freedom are $n - \text{tr}(\mathbf{S}) = 312 -$

---

[7]Depending on the software used, confidence bands may or may not be automatically estimated by the nonparametric regression routine.

$5.14 = 306.86$. Using the formula from above, the calculation is

$$F = \frac{(25\,684.06 - 21\,811.94)/(5.14 - 1)}{21\,811.94/306.86} = 13.17. \qquad (2.24)$$

This test statistic is F-distributed with 4.14 and 306.86 degrees of freedom. The resulting p-value is <0.001, which means we can reject the null hypothesis that there is no relationship between the challengers' vote share and support for Perot. While this hypothesis test is useful, a linear regression would have told us much the same thing. A more useful hypothesis test is one that compares the non-parametric fit to a globally linear fit. That is we want to test whether there is a 'significant' amount of nonlinearity. To perform this test, we compare the fit of linear model to the fit of *loess* model. The residual sum of squares for the linear model is 23 484.27 and J is $5.14 - 2$. That is the *loess* fit uses 5.14 equivalent parameters and the linear model uses two, one for the constant and another for the slope. The test statistic is

$$F = \frac{(23\,484.27 - 21\,811.94/(5.14 - 2)}{21\,811.94/306.86} = 7.50. \qquad (2.25)$$

The above test statistics is F-distributed with 3.14 and 306.86 degrees of freedom. For this test statistic, $p < 0.001$, which implies that there is a significantly nonlinear relationship between challengers' vote share and support for Perot. The ability to test the LRP model against the linear regression model provides analysts with a powerful tool for the diagnosis of nonlinear functional forms. At a minimum, the test indicates that a model of the relationship between these two variable requires a power transformation. A linear model that does not take this nonlinearity into account will be misspecified and provide biased and possibly misleading estimates.

Given that the functional form appears to be nonlinear, an analyst might choose to log or square the measure of Perot support to better capture the nonlinear functional form. Unfortunately, the use of either power transformation again requires the use of a functional form assumption that is difficult to evaluate *prima facie*. The nonparametric regression model, however, allows us to test the adequacy of power transformations just as we tested the adequacy of the linear fit. For example, we fit a model with a quadratic term for Perot support. The residual sum of squares for this model is 22 334.06. Therefore, the test statistic to test the quadratic model against the *loess* model is

$$F = \frac{(22\,334.06 - 21\,811.94/(5.14 - 3)}{21\,811.94/306.86} = 3.44 \qquad (2.26)$$

with 2.14 and 306.86 degrees of freedom, we find that $p = 0.03$. A quadratic power transformation, then, fails to adequately account for the nonlinearity in the model. We also performed the same test comparing the *loess* model with one using a logarithmic transformation. For this test, $p = 0.013$, so a regression model with a power transformation fails to account for the nonlinearity. One could fit the model with other power transformations and test them in a similar fashion or simply use the nonparametric fit.

2.5   Multiple Nonparametric Regression

Thus far, we have restricted our attention to bivariate relationships. Of course, most applied research in the social sciences considers more complicated multivariate specifications. Local polynomial nonparametric regression models also generalize to a multivariate framework. As such one can estimate the following nonparametric regression model

$$y_i = f(x_{1i}, x_{2i}, \ldots, x_{ki}) + \varepsilon. \tag{2.27}$$

Estimation of the above model proceeds in similar fashion to that of a bivariate nonparametric model. Instead of a bivariate local polynomial regression, the local estimate is based on multiple regression. As before, data that are farther away from the focal point receive less weight than data near the focal point. The definition of the local bin width is more complex since the calculation of the bin size becomes the distance along a plane. The predicted value from the local multivariate polynomial regression model becomes the local estimate that is connected to form a plane which represents the joint effect of x_{1i} and x_{2i} on y. Importantly, the bivariate nonparametric regression model is equivalent to the specification of a nonlinear interaction model. The resulting estimate will be a plane that we can plot as the joint effect of the two predictor variables on y.

We return to the data on the 1992 House elections for an example of multivariate nonparametric regression. While Jacobson and Dimock (1994) examine the relationship between support for Perot and challengers, they also try to isolate whether the House Bank scandal also contributed to the strong showing by challengers in 1992. In the House Bank scandal, it was found that members of the House were often overdrawing their accounts at the House Bank. To measure the effect of the House Bank scandal, Jacobson and Dimock use a measure that records the number of overdrafts by each member of the House. They also include an interaction between the number of overdrafts and the amount spent by the challenger during the election. We estimate a multivariate model of challenger's vote share as a function of the number of overdrafts by a House member

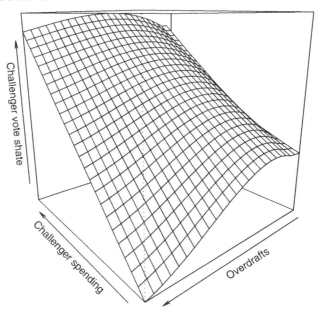

Figure 2.14   The joint effect of overdrafts and challenger spending on the challenger's vote share.

and challenger spending. We use *loess* with a span parameter of 0.8 to model the joint effect of challenger spending and the number of overdrafts. Figure 2.14 contains the estimates from this model.

One unique aspect of these models is that the estimate depicts the joint effect of overdrafts and spending. The model output is a plot of the interactive effect of the two variables on the outcome variable. The multivariate nonparametric regression model provides a means for the modeling of nonlinear interaction effects. It is apparent in Figure 2.14 that the effect of overdrafts is contingent on the level of challenger spending. Even if an incumbent overdrafted in a profligate manner, it helped the challenger little unless challenger spending was also increasing. Importantly, multivariate nonparametric models provide the analyst with a means of graphically representing interaction terms. We can also test whether adding the spending variable improves the model fit using the same inferential techniques based on the F-test. The results from such a test clearly indicate that including challenger spending in the model improves the fit as the resulting p-value is well below the standard 0.05 threshold ($p < 0.001$).

Unfortunately, these models are not without their flaws. The first problem is the 'curse of dimensionality'. As the number of predictors in the model increase, the number of observations in the local neighborhood of the focal point tends to decline rapidly. One solution is to increase the span, but increasing the span increases the bias in the estimates. Secondly, interpretation becomes impossible with more than two predictors. These models do not return any parameters, and when more than two predictors are included in the model, visual representation of the effect is not possible. Readers further interested in the details of multivariate smoothing should consult Wood (2006). In Chapter 5, we will explore methods for extending nonparametric models to multiple predictors.

2.6   Conclusion

In this chapter, we have explored the basic ideas behind nonparametric regression. The local polynomial regression model provides a flexible and powerful means of estimating nonlinear relationships. Much of the power of these models stems from the ability to test the nonparametric fit against the linear model or other nonlinear but parametric models. The LPR model can also be extended to multivariate settings. While there are serious limitations in multivariate settings, these model do provide a unique means of exploring interactive effects. However, care must be taken not to overfit the data by setting the span parameter to small. Moreover, analysts must take care not to overinterpret instances of local nonlinearity. In the next chapter, we present a different form of smoother: the spline. While in many settings, spline smoothers do not offer much beyond LPR, penalized splines have several attractive properties that make them superior to LPR. Moreover, splines are more readily incorporated into the semiparametric regression model.

2.7   Exercises

Data sets for exercises may be found at the following website: http://www.wiley.com/go/keele_semiparametric.

1. The data in cps.dta is a subset of Current Population Survey data from 1996. The dependent variable is a measure of hourly wages. The data set also contains two independent variables: age and income from interest and investments. Use this data for the exercise below.

   (a) First, use a scatterplot to study the relationship between wage and age. Does their appear to be a statistical relationship between the two variables? Does it appear to be linear or nonlinear?

(b) Sort the data by age, and then calculate the average level of wages for 10 different age groups. Plot the means and connect them with a line. Does the relationship appear to be nonlinear?

(c) Use software to program a moving average smoother and again plot the relationship between wages and age. Adjust the bandwidth for different fits. Is there enough data to smooth within each age category?

(d) Estimate a kernel smoother, trying the tricube, Gaussian, and box kernels. Do the various kernels change your inference about the relationship between age and wages?

(e) Use local polynomial regression to estimate the relationship between age and wages. Set the span using the visual method. At what span value does the relationship between age and wages appear to be overfit?

(f) Calculate 95% confidence interval bands and plot the relationship with CI bands for a *loess* fit between wage and age.

(g) Test whether the relationship is statistically significant and whether there is a significant amount of nonlinearity. Is the nonparametric fit better than either a logarithmic or quadratic transformation of age?

(h) Finally, add interest income to the model and plot the joint nonparametric effect. Does interest income improve the fit of the model?

2. The `palmbeach.dta` data set contains precinct level voting data for the 2000 election in Florida. The data set contains an indictor for each precinct (of which there are 213), the number of votes that Al Gore received in each precinct, and the number of overvotes reported in each precinct. A ballot is labeled as an overvote when a voting counting machine records votes for more than one candidate for a single office. Repeat the steps from above for the relationship between the votes for Al Gore and the number of overvotes in each precinct. You need not smooth in multiple dimensions for this problem.

3

Splines

This chapter provides coverage of spline smoothers. Spline smoothers are another nonparametric regression technique used with scatterplots. One might question whether another smoothing technique is necessary given local polynomial nonparametric regression (LPR) models. In truth, for basic scatterplot smoothing, often one cannot tell much difference between spline and LPR fits. Splines, however, have several advantages over local polynomial regression. First, splines have an analytic foundation that is superior to that of local regression, as one can prove that a spline smoother will provide the best mean squared error fit. Second, one type of spline, the smoothing spline, is designed to prevent overfitting, a prominent concern with nonparametric smoothers. Third, there have been a number of advances in the methods used to estimate splines, while advances in local regression has been fairly static. As a result, the software for fitting spline models is typically superior to that for local regression. For example, most implementations of spline produce confidence bands, which may not be true for LPR smoothers. Moreover, other differences in the estimation algorithms will be obvious in the next chapter on automated smoothing. Finally, splines are easier to incorporate into semiparametric estimation, and they have become the smoothing method of choice for semiparametric regression models.

3.1 Simple Regression Splines

The term 'spline' originally referred to a tool used by draftsmen to draw curves. For our purposes, splines are piecewise regression functions we constrain to join at points called knots. In their simplest form, splines are regression models with a set of dummy variables on the right hand side of the model that we use to force the regression line to change direction at some point along the range of X. For the simplest regression splines, the piecewise functions are linear, a constraint that we will later relax. In essence, we fit separate regression lines within the regions between the knots, and the knots tie together the piecewise regression fits. Again, splines are a local model with local fits between the knots instead of within bins that allow us to estimate the functional form from the data.

Like local polynomial regression, the analyst must make several modeling decisions with splines. For LPR, we had to choose the degree of the polynomial, the span, and perhaps the weighting function. With splines, one must choose the degree of a polynomial for the piecewise regression functions, the number of knots, and the location of the knots. In Chapter 2, the choice of the span parameter, the percentage of data used in each local fit, proved to be critical since it affected the smoothness of the nonparametric estimate, while the other choices were less important. With splines, we again find that while the fit is invariant to some of the modeling choices, the analyst must focus on how smooth the fit should be. For some types of splines, the number of knots will control the amount of smoothing, while for other types of splines, a smoothing parameter controls the smoothing.

Perhaps the most confusing aspect of splines is that there are so many different types. For example, there are regression splines, cubic splines, B-splines, P-splines, natural splines, thin-plate splines, and smoothing splines to name a few. Moreover, there are often combinations such as natural cubic B-splines. The wide variety of splines partially stems from the progress in research on splines. Often a new type of spline either supplants an older type of spline or adds a refinement to existing methods. The splines most often used in statistics, smoothing splines, are more complex than regression splines, but they work on the same principle. Thus, in this next section, we focus on understanding the simplest type of splines before moving to more complicated types.

We start with the basic equation for a smooth fit between x and y:

$$y = f(x) + \varepsilon. \tag{3.1}$$

As before, we only assume that f is some smooth function, but we would like to estimate this model using a single regression model estimated with ordinary least squares as opposed to using a series of local models. To do this, we need to represent the model matrix so that we can estimate Equation (3.1)

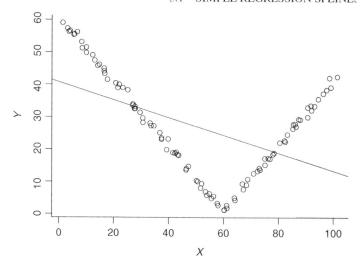

Figure 3.1   A simulated nonlinear functional form with OLS fit.

with least squares. This process is best demonstrated with a simple illustration. Figure 3.1 contains a scatterplot of the nonlinear relationship between two simulated x and y variables. The form of the nonlinear dependency between x and y is is quite obvious. The solid line in Figure 3.1 represents the estimate from a global linear fit between the two variables. The global estimate clearly fails to capture the nonlinear dependency between x and y, as it only indicates that there is a fairly strong negative relationship between the two variables.

The logic behind a regression spline is to estimate two separate regression lines that will be joined at the kink in the data. The first regression line will approximate the negative dependency between the two variables, and the second regression line will approximate the upturn in the functional form. In this instance, we can clearly identify the change point in the relationship between x and y, and therefore, we can easily use two piecewise linear estimates joined at the change point. To estimate the spline model, we need to specify the point where the two separate OLS fits will be joined. Placement of the joinpoint, or knot, between the two regression lines here is easy since the kink is so obvious. We only need a single knot since there are just two piecewise linear fits to conjoin. Additional piecewise fits would require additional knots. We denote the single knot with c_1. Using c_1, we can write the following regression spline model:

$$y = \alpha + \beta_1 x + \beta_2(x)_+ + \varepsilon \qquad (3.2)$$

where

$$(x)_+ = \begin{cases} x & \text{if } x > c_1 \\ 0 & \text{if } x \leq c_1. \end{cases} \tag{3.3}$$

The $(\cdot)_+$ function indicates that $(x)_+$ equals x if x is greater than the knot value and is equal to 0 otherwise. If we assume that the function is piecewise linear, we can rewrite Equation (3.2) as two separate but conjoined regression lines:

$$y = \begin{cases} \alpha + \beta_1 x & \text{if } x \leq c_1 \\ \alpha + \beta_1 x + \beta_2 (x - c_1) & \text{if } x > c_1. \end{cases} \tag{3.4}$$

We have now rewritten a single linear model as two separate linear fits that describe each part of the nonlinear dependency between x and y. To jointly estimate the two piecewise regression fits requires a set of basis functions. Basis functions are an important concept in the estimation of splines, and we spend some time explaining basis functions for splines before estimating the current example.

3.1.1 Basis Functions

In linear algebra, the basis of a vector space (a set of vectors) is the number of columns or rows of that vector space that can be expressed as a linear combination. In the context of regression models, $\mathbf{X}$, the model matrix, is a vector space with a corresponding basis function. For example, if we have a regression model with a constant and a single covariate, the corresponding basis functions for this model would be a vector of 1's and the vector x_1, the lone predictor variable. In fact, the right hand side of any regression model is a linear combination of basis functions and therefore forms a basis. We use additional basis functions to approximate nonlinearity for regression splines. The simplest way to add to the basis is to use additional predictor variables. With splines, we add another basis to the data matrix, but this basis is a transformation of the single predictor. These additional basis function will allow us to approximate the nonlinear relationship between x and y. To add an additional basis requires a set of basis functions, one basis function for each piecewise function. In the current example, we have two piecewise functions to estimate, so we must define two basis functions for the two piecewise linear fits, one for the left side knot and one for the right side of the knot:

$$B_L(X) = \begin{cases} c - x & \text{if } x < c \\ 0 & \text{otherwise} \end{cases} \tag{3.5}$$

$$B_R(X) = \begin{cases} x - c & \text{if } x > c \\ 0 & \text{otherwise.} \end{cases} \quad (3.6)$$

These basis functions are applied to the model matrix to form a new basis. In practical terms, application of the basis functions adds another vector to the data matrix. Before, the model matrix was comprised of a constant and x. The model matrix for the spline model will now consist of a constant and two data vectors. The first data vector has values of c_1 less x until the value of the knot equals the x values and then is zero for the rest of the values of x. The second data vector in the model matrix consists of zeros until we get to the value of x where we have placed a knot, and then the vector will take values of x less the knot value. Applying the basis functions adds an additional basis, before the model had a basis of dimension 2 and now will have a basis of dimension 3. In essence, by applying the basis functions, we have added an additional regressor to the model matrix. For the current example, to apply the basis functions, we must choose where to place the knot. We place a single knot at 60 since in the simulated data this the exact x value where the kink occurs. This is the point in x where we conjoin the two piecewise linear fits. With the knot value chosen, we apply the basis functions to the simulated data. Practical implementation of the basis functions consists of writing two functions in a programming environment, applying those functions to the simulated x variable and then constructing a new model matrix comprised of a constant and the results from the application of each basis function. Below is an example of the model matrix, $\mathbf{X}$, constructed after we applied the two basis functions to the simulated x variable.

$$\mathbf{X} = \begin{bmatrix} 1 & (60 - x_1) & 0 \\ \vdots & \vdots & \vdots \\ 1 & (60 - x_{59}) & 0 \\ 1 & 0 & 0 \\ 1 & 0 & (x_{61} - 60) \\ \vdots & \vdots & \vdots \\ 1 & 0 & (x_n - 60) \end{bmatrix}. \quad (3.7)$$

Once we have formed a new model matrix, estimating a spline fit between x and y is simple. We use the new model matrix to construct the usual 'hat' matrix: $(\mathbf{X}'\mathbf{X})^{-1}\mathbf{X}'$. Application of the hat matrix to the data vector for the outcome produces a set of predictions that form the nonparametric spline estimate of the relationship between x and y. Therefore, the spline estimate is

$$\hat{\mathbf{y}} = \mathbf{H}\mathbf{y} \quad (3.8)$$

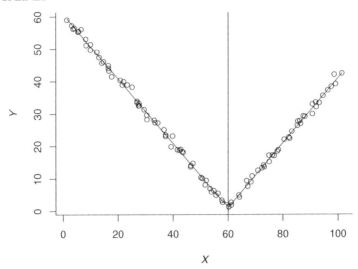

Figure 3.2   A piecewise linear regression spline fit.

where $\mathbf{H} = (\mathbf{X}'\mathbf{X})^{-1}\mathbf{X}'$. We then plot the predictions from the model to view the spline estimate. Figure 3.2 contains a plot of the resulting estimated spline fit between x and y, which closely approximates the nonlinear relationship between x and y.

This simple example nicely illustrates the basic procedure for using splines. It allows the reader to see that the regression spline estimate results from simply applying least squares to a model matrix altered with the basis functions. We can use a combination of additional knots and more complex basis functions to approximate more complex nonlinear relationships. Clearly much depends on knot placement, a topic we will take up shortly, but since splines models are essentially transformed regression models, it allows for an easy mixing of nonparametric estimation with more standard statistical models. One drawback to the simple regression splines we used in the current example is the restrictive assumption of piecewise linear functions, but we can easily relax this assumption.

3.2   Other Spline Models and Bases

Statisticians have developed a wide variety of basis functions for spline models. In general, a change of basis will not change the fit between x and y very much. Typically, different spline bases are used for better numerical stability and ease of implementation (Ruppert, Wand, and Carroll 2003). In this section, we review some of the more commonly used bases for spline models as well as a variety improvements that have been proposed for splines.

3.2.1   Quadratic and Cubic Spline Bases

The simple regression splines we used in the last section to estimate the nonlinear dependence between the simulated x and y are not suitable for most applied smoothing problems. It is overly restrictive to only estimate piecewise functions that are linear between the knots since we wish to estimate more curvilinear functional forms. The solution is to combine piecewise regression functions with polynomial regression by representing each piecewise regression function as a piecewise polynomial regression function. Piecewise polynomials offer two advantages. First, piecewise polynomials allow for nonlinearity between the knots. Second, piecewise polynomial regression functions ensure that the first derivatives are defined at the knots, which guarantees that the spline estimate will not have sharp corners.[1]

Altering the regression splines used thus far to accommodate piecewise polynomials is simple. For the spline model in the last section, we could estimate piecewise polynominal fits by adding x^2 to the basis and squaring the results from the basis functions. These alterations form a quadratic spline basis with a single knot at c_1. Typically, cubic spline bases are used instead of quadratic bases to allow for more flexibility in fitting peaks and valleys in the data. A spline model with a cubic basis and two knots c_1 and c_2 is formed from the following linear regression model:

$$Y = \alpha + \beta_1 x + \beta_2 x^2 + \beta_3 x^3 + \beta_4(x - c_1)_+^3 + \beta_5(x - c_2)_+^3 + \varepsilon. \qquad (3.9)$$

The spline estimate is again the predictions from the hat matrix applied to the outcome variable. To form the hat matrix, we must first construct a model matrix that contains the correct bases. For this example, the model will contain the following constructed data vectors:

$$x_1 = x$$
$$x_2 = x^2$$
$$x_3 = x^3$$
$$x_4 = (x - c_1)_+^3$$
$$x_5 = (x - c_2)_+^3 \qquad (3.10)$$

---

[1]Recall that for a continuous function to be differentiable at a particular point, the function must not change dramatically at that point. Sharp corners in functions are places where the first derivative is not defined. For example, for the function $y = |x|$ the derivative does not exist at $x = 0$. For polynomial functions, such sharp corners, where the first derivative is undefined, do not exist.

where x represents the original predictor variable. The model matrix will consist of a constant and the above five variables. We use this model matrix to form a hat matrix that is applied to the outcome variable, and the predictions from this model serve as the spline estimate of the possibly nonlinear relationship between x and y. The number of parameters used to construct the spline estimate is controlled by the number of knots. If there are k knots, with a cubic basis, the function will require $k + 4$ regression coefficients (including the intercept). The cubic basis allows for flexible fits to nonlinearity between the knots and eliminates any sharp corners in the resulting estimate. The later is true since the first derivative exists for $(x - c_1)^3_+$ and it follows that the first derivative will also exist for any linear combination of the terms in 3.10 (Ruppert, Wand, and Carroll 2003). For cubic regression splines, there are a number of equivalent ways to write the basis. For example, below we outline another way to write the cubic basis that is very convenient for estimation but looks rather daunting (Gu 2002). First, let the knot locations in x be denoted by x^*. We will have to select these knot locations, typically they are evenly spaced over the range of x. Now, we define $B(x, x^*)$ as the following function to represent the basis:

$$B(x, x^*) = [(x^* - 1/2)^2 - 1/12][(x - 1/2)^2 - 1/12]/4$$
$$-[(|x - x^*| - 1/2)^4 - 1/2(|x - x^*| - 1/2)^2 + 7/240]/24. \quad (3.11)$$

Applying the above function to x does most of the work in the construction of a model matrix that allows for f to be estimated with a linear regression model. Application of the above function and appending the result to a constant and the x vector produces a model matrix where the ith row is

$$\mathbf{X}_i = [1, x_i, B(x_i, x_1^*), B(x_i, x_2^*), \ldots, B(x_i, x_{q-2}^*)]. \quad (3.12)$$

The form of the model matrix above is identical to past discussions of the basis. The first column of the data matrix is for the constant, the next is the x variable, and the following columns form a set of piecewise cubic functions that are fit between each knot location. We can use a data matrix to form the hat matrix and estimate the spline fit for the relationship between x and y. The number of knots selected by the analyst will determine the dimension for the basis.

    We now illustrate the use of cubic regression splines. For comparability, we return to the example from the 1992 House elections to again estimate the relationship between support for the challenger and support for Perot. In Chapter 2, we used local polynomial regression to fit a nonparametric regression model to these two variables. The result revealed a nonlinear dependence between these two variables due to a threshold effect. We now use cubic splines to estimate a nonparametric regression model for these same two variables.

For the analysis, we used a series of user-written functions instead of a pre-programmed function from a statistical software package. These functions are available on the book website and help make the basic operation of splines readily transparent. First, we rescale the Perot support variable to lie in the [0, 1] interval. This rescaling of the variable allows for easier placement of the knots. Next, we select the number of knots; for this model, we use four knots, which implies a rank 6 basis or that there are six columns in the model matrix.[2] How does the number of knots affect the spline estimate? The number of knots determines the smoothness of the nonparametric estimate. Increasing the number of knots increases the number of piecewise functions to produce a more flexible fit. Using two knots produces a globally linear fit. We selected four knots on the basis of a visual trial and error method. We must also decide where to place the four knots. We place the knots at equal intervals over the range of the x variable starting at an x value of 0.2, which implies that the other three knots are placed at 0.4, 0.6, and 0.8. We now construct the model matrix. The first column in the model matrix is a vector of 1's, the second column is the vector x, the support for Perot variable. Given that there are four knots there will be four additional vectors in the model matrix. To create these data vectors, we apply the function in Equation (3.11) to x within each knot placement. Once the model matrix is completed, the process is simple. We form a hat matrix based on the constructed model matrix and apply it to the challenger's vote share to form the model predictions which make up the spline nonparametric estimate. We plot the spline estimate in Figure 3.3. The resulting spline looks very similar to the results using *lowess* in Chapter 2. We see the same nonlinear functional form where support for the challenger no longer increases once a threshold of support for Perot is reached.

This example demonstrates one advantage of spline smoothers: their simplicity. In this example, we estimated a nonlinear functional form with nothing more complicated than ordinary least squares and a transformed model matrix. While this simplicity matters less for nonparametric estimates, it will be valuable once we consider semiparametric estimation. The next type of spline smoother is not a change in basis but simply a convenient alteration of cubic splines to produce a better fit.

3.2.2   Natural Splines

While cubic splines are widely used, they are often altered slightly to improve the fit. One limitation of cubic splines is that the piecewise functions are only fit

---

[2]The number of vectors that span the basis is referred to as the rank.

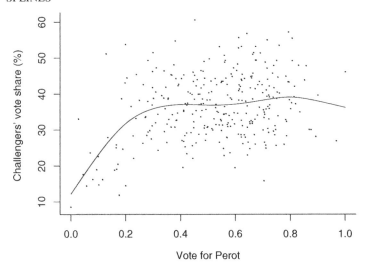

Figure 3.3   Support for the challenger as a function of support for Perot in the 1992 House elections.

between each knot. For data that falls before the first knot and beyond the last knot, we do fit not a piecewise function. Without fits to the boundary of the data, it is possible for the spline fit to behave erratically around the limits of x. Natural cubic splines add two knots to the fit at the minimum and maximum values of x and fit a linear function between the additional knots at the boundary and the interior knots. This constrains the spline fit to be linear before the first knot and after the last knot. Such enforced linearity at the boundaries avoids any wild behavior in the spline fit near the extremes of the data. Cubic splines may not display erratic fits at the boundaries, but natural splines can improve the overall spline fit should problems occur. Since little is lost by using natural splines while some gains in model fit are possible, natural cubic splines are generally preferred to cubic splines. Later in the chapter, we compare natural cubic splines to standard cubic splines.

3.2.3   B-splines

There is one further refinement that is also typically applied to cubic splines. For cubic splines (natural or otherwise), the columns of $\mathbf{X}$, the model matrix, tend to be highly correlated since each column is a transformed version of x, which can induce considerable collinearity. The collinearity may result in a nearly singular model matrix and imprecision in the spline fit (Ruppert, Wand, and Carroll 2003).

As a remedy, one can represent the cubic spline (and any other polynomial basis) as a B-spline basis. A k-knot cubic B-spline basis can be represented as:

$$f(x) = \sum_{i=1}^{k} B_i^2(x)\beta_i \qquad (3.13)$$

where the B-spline basis functions are defined as:

$$B_i^2(x) = \frac{x - c_i}{c_{i+2+1} - c_i} B_i^{2-1}(x) + \frac{c_{i+2+1} - x}{c_{i+2+1} - c_{i+1}} B_{i+1}^{2-1}(x) \ i = 1, \ldots k \qquad (3.14)$$

and

$$B_i^{-1}(x) = \begin{cases} 1 & \text{if } c_i \leq x < c_{i+1} \\ 0 & \text{otherwise.} \end{cases} \qquad (3.15)$$

The B-spline basis function is, in essence, a rescaling of each of the piecewise functions. The idea is similar to rescaling a set of X variables by mean subtraction to reduce collinearity. The rescaling in the B-spline basis reduces the collinearity in the basis functions of the model matrix. The resulting spline model is more numerically stable than the cubic spline. This is especially true if one is using a large number of knots and OLS is used to fit the spline model. See de Boor (1978) and Eilers and Marx (1996) for details. Many spline routines contained in statistical software are cubic B-splines or natural cubic B-splines. Before taking up another example to demonstrate the performance of the spline models outlined thus far, we take up the topic of knots.

3.2.4   Knot Placement and Numbers

In any spline model, the analyst must select the number of knots and decide where they should be placed along the range of x. In the example with simulated data, deciding on the placement of the single knot was trivial given the piecewise linearity of the function. Visual examination of the scatterplot usually, however, does little to aid the placement of knots. For example, the scatterplot between the challenger vote share and support for Perot reveals little about where one might place the knots. The knot placement we used for the spline estimate in Figure 3.3 probably appeared to be arbitrary. Knots tend to generate considerable confusion, so in this section we focus on both how to place knots and how many knots to use.

Knot placement is, however, not a complicated model choice. Stone (1986) found that where the knots are placed matters less than how many knots are used.

Standard practice is to place knots at evenly spaced intervals in the data. Equally spaced intervals ensure that there is enough data with each region of x to get a smooth fit, and most software packages place the knots at either quartiles or quintiles in the data by default. The analyst can usually override default placement of knots. If the data have an obvious feature, it may be useful to place the knots in a less automatic fashion.

But the question of how to select the number of knots remains, and the number of the knots has an important effect on the spline fit. The number of knots chosen affects the amount of smoothing applied to the data by controlling the number of piecewise fits. A spline with two knots will be linear and globally smooth since there is only one piecewise function. Increasing the number of knots increases the number of piecewise functions fit to the data allowing for greater flexibility. More piecewise functions results in an increased number of local fits. If one selects a large enough number of knots the spline model will interpolate between the data points, since more knots shrink the amount of data used for each piecewise function. The number of knots effectively acts as a span parameter for splines. Therefore, one is faced with the same tradeoffs embodied in the choice of the span parameter. If one uses a small number of knots, the spline estimate will be overly smooth with little variability but may be biased. Using a high number of knots, conversely implies little bias but increases variability in the fit and may result in overfitting.

Understanding how the number of knots affects the spline fit does not help in knowing how many to use. Fortunately, the spline fit is usually not overly sensitive to the number of knots selected, and two different methods exist for knot number selection. One method is to use a visual trial and error process as we did for span selection. Four knots is the standard starting point. If the fit appears rough, knots are added. If the fit appears overly nonlinear, knots are subtracted. Four or five knots is sufficient for most applications. The number of knots is also loosely dependent on sample size. For sample sizes above 100, five knots typically provides a good compromise between flexibility and overall fit. For smaller samples, say below 30, three knots is a good starting point.

The second method is less *ad hoc* than the visual method. Since each knot represents additional parameters being added to the model, Eilers and Marx (1996) recommend using Akaike's Information Criterion (AIC) to select the number of knots. One chooses the number of knots that returns the lowest AIC value. Using the AIC is less arbitrary than the visual method and produces reasonable results. As we will see later in the chapter, newer spline models dispense with choices about knots. Now, we return to the Congressional elections example to better understand the performance of the spline models discussed thus far and to explore how the number of knots affects the final estimate.

3.2.5   Comparing Spline Models

As before, we are interested in observing whether there is a nonlinear relationship between the challenger's vote share and support for Perot. Thus far, LPR and simple cubic spline estimates have revealed a nonlinear dependency between these two variables. We now use both cubic B-splines and natural cubic B-splines to estimate the nonparametric fit between these two variables. The software used in this example places the knots at the quantiles based on the distribution of the data by default. There do not appear to be obvious locations for the knots, so we rely on the software defaults. We use four knots for both spline models, and two knots are added at the minimum and maximum values of the Perot support variable for the natural spline model. We will evaluate whether this is the optimal number of knots in a moment. Both spline estimates are in Figure 3.4.

We see little difference between the cubic B-splines and the natural cubic B-splines estimates. In both, the by now familiar threshold effect is present, where the challenger's vote share levels off once support for Perot exceeds about 15% in the district. One might argue that the two spline models do differ slightly at the end points of each fit. The cubic fit curves up more for the higher values of support for Perot, while the natural spline fit displays less curvature since it has been constrained to be linear at the end points. As noted previously, with cubic splines strange behavior is possible at the extremes of x, so in general, natural splines are preferred.

In both of the spline models in Figure 3.4, we used the rule-of-thumb starting point of four knots. A useful exercise is to explore the effect of using different numbers of knots. We could simply add and subtract knots from the four knot baseline and observe whether the fit changes, but instead we calculate the AIC for a range of knot numbers. The lowest AIC value provides a summary measure of the best fit for the fewest number of knots. Using natural cubic B-splines, we

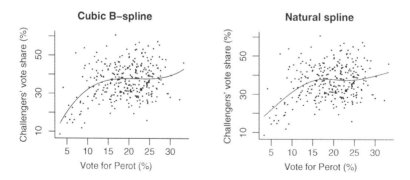

Figure 3.4   Cubic B-Spline and natural spline fit to challenger's vote share.

Table 3.1   AIC values for differing numbers of knots.

2 knots	2239.6
3 knots	2228.0
4 knots	2223.4
5 knots	2220.4
6 knots	2221.0
7 knots	2222.6
8 knots	2224.5
9 knots	2225.8

estimated models with two–nine knots and calculated the AIC for each model. The calculated AIC values are in Table 3.1.

The spline fit with five knots returns the lowest AIC value, which indicates that we would prefer a model with five knots. Using more than two knots always produces a substantially better fit than a model with two knots. This is preliminary evidence that a linear fit provides an inadequate description of the data. Figure 3.5 contains plots for the natural spline fits with four, five, six, and nine knots. While the AIC indicates that we should use five knots, there is little difference across the four estimates. The reader should notice that for the model with nine knots, we observe excess fit, as idiosyncratic variation is evident, which is undoubtedly caused by overfitting. Despite the overfitting, the same basic relationship is observed between the two variables. It is often the case that analysts worry that spline models are overly sensitive to knot selection, but as the example demonstrates, the fit is largely invariant to the number of knots used. While too many knots will cause overfitting and induce small and presumably random variation in the fit, the AIC provides a clear and familiar criterion for selecting the number of knots. We now turn to smoothing splines which are designed to further limit overfitting.

3.3   Splines and Overfitting

If we want a flexible estimate of the statistical relationship between two variables, both splines and local polynomial regression can provide such an estimate with few assumptions about the functional form. A common criticism of both of these nonparametric regression models in the social sciences is that they are too flexible. The concern with both methods is that it is easy to have a surfeit of (local) parameters, which produces overly nonlinear nonparametric estimates that overfit data. Critics argue that while nonparametric regression estimators make few

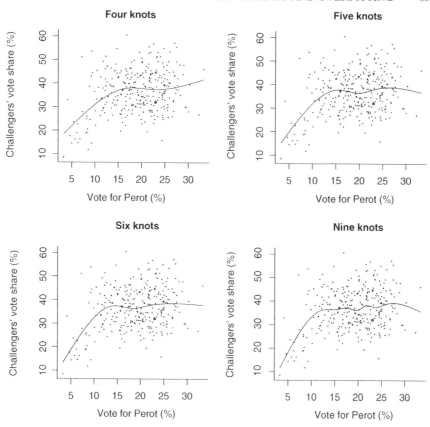

Figure 3.5   Differing numbers of knots with natural splines.

assumptions, they too easily display idiosyncratic local variation between x and y that is not of substantive interest. In short, instead of observing an estimate of the possibly nonlinear functional form, the analyst observes a nonlinear estimate due to overfitting.

While overfitting is possible with nonparametric regression, many of these criticisms are overstated. It is often the case that an analyst would have to use a very small span setting or a large number of knots to overfit the data. This has been the case in the Congressional election example as it required a span setting of 0.20, or nine knots to overfit the data. That said, it is possible to overfit relationships using nonparametric regression. The standard remedy for such overfitting has been a suggestion that analysts err on the side of undersmoothing when choosing either the span or the number of knots. While undersmoothing provides a simple solution to the problem of overfitting, it is not a solution that appeals to any principles of statistical theory. Moreover, this is a solution that suggests erring on the side of increasing the bias in the fit. We would prefer a less *ad hoc* solution grounded in statistical theory. Penalized splines are a nonparametic regression technique that relies on principles of statistical theory to minimize the possibility of overfitting.

3.3.1   Smoothing Splines

Of course, it is possible to overfit both parameteric and nonparametric regression models. Overfit statistical models have too many parameters relative to the amount of data and cause random variation in the data to appear as a systematic effects. The solution to overfitting is to reduce the number of parameters in the model, but the parameters cannot be dropped in an *ad hoc* fashion. The solution is to reduce the parameters subject to some criterion. The AIC is one criterion for reducing parameters, and as we saw in the last section can be applied to splines.

Another solution to overfitting is penalized estimation. Here, for each parameter used in the model, a penalty is added to the model. Penalized estimation is not widely used in the social sciences but is relatively common in statistics. Social scientists are, however, very familiar with one statistic that relies on the principles of penalized estimation: the adjusted R^2. The adjusted R^2 measures the fit of a model but does so subject to a penalty for each additional parameter used in the model. Smoothing splines operate in a similar fashion by placing a penalty on the number of local parameters used to estimate the nonparametric fit.

Since spline models are estimated with least squares, they share the properties of linear regression models. This implies that the spline estimate, $\hat{f}$, minimizes the sum of squares between y and the nonparametric estimate, $f(x_i)$

$$\mathrm{SS}(f) = \sum_{i=1}^{n} [y - f(x)]^2. \tag{3.16}$$

The concern, however, is that the estimate of f that minimizes Equation (3.16) may use too many parameters. The penalized estimation solution is to attach a penalty for the number of parameters used to estimate f. This suggests that we minimize $\mathrm{SS}(f)$ but subject to a constraint or penalty term for the number of local parameters used. The penalty we use for spline models is

$$\lambda \int_{x_1}^{x_n} [f''(x)]^2 dx \tag{3.17}$$

which implies that the spline estimate is

$$SS(f, \lambda) = \sum_{i=1}^{n} [y - f(x)]^2 + \lambda \int_{x_1}^{x_n} [f''(x)]^2 dx. \tag{3.18}$$

In Equation (3.18), we minimize the sum of squares between y and the nonparametric estimate, $f(x)$ subject to the penalty in Equation (3.17).

The term in Equation (3.17) is the constraint known as a roughness penalty. This penalty has two parts. The first is λ, often referred to as the smoothing or tuning parameter, and the second is the integrated squared second derivative of $f(x)$. The logic behind the use of the integrated squared second derivative of $f(x)$ is fairly intuitive. The second derivative measures the rate of change of the slope for a function or curvature. A large value for the second derivative means high curvature and vice versa.[3] Through the use of the squared integral, the term sums a measure of curvature along the entire range of the nonparametric estimate in essence giving us a measure of curvature along the range of the nonparametric estimate. When it is large, $f(x)$ is rougher, and when it is small, $f(x)$ is smoother.

The λ parameter, which is nonnegative, directly controls the amount of weight given to the second derivative measure of how smooth f is. Therefore, λ establishes a tradeoff between closeness of fit to the data and the penalty giving λ a function analogous to the bandwidth for a *lowess* smoother. As the value for λ decreases, then $\hat{f}(x)$ interpolates the data, and we get a rough fit. As $\lambda \longrightarrow \infty$ the integrated second derivative is constrained to be zero, and the result is the smooth global least squares fit. More generally, as λ increases, we get a smoother fit to the data but perhaps a biased fit, and as the parameter decreases, the fit displays less bias but with increased variance.

While a very small value for λ comes close to interpolating the data and a large value of λ returns a least square fit, intermediate values of λ often do not have an interpretable effect on the amount of smoothing applied to the data. To solve this problem, we can transform λ to be an approximation of the degrees of freedom, which is logical since we might view λ as controlling the number of local parameters used for the nonparametric estimate. The degrees of freedom for smoothing splines are calculated in a manner similar to that for OLS models and local polynomial regression, but the penalty creates some complications. Recall from the last chapter that the degrees of freedom for a linear regression model is equal to the number of fitted parameters which is equal to $\text{tr}(\mathbf{H})$ where $\mathbf{H} = \mathbf{X}(\mathbf{X}'\mathbf{X})^{-1}\mathbf{X}$. For standard spline models, which rely on an altered model matrix

---

[3]As another example, the Hessian is a matrix of second derivatives that measures the amount of curvature around the likelihood maximum.

but are then fit with least squares, the degrees of freedom would also be calculated by tr($\mathbf{H}$). To calculate the degrees of freedom for penalized splines, we must generalize $\mathbf{H}$.

First, we write the penalized spline model in Equation (3.18) in matrix form. Given the equivalence between splines models and linear regression models, we can write the first term in Equation (3.18) as linear regression model in matrix form. It can be shown that the penalty term from Equation (3.18) can be written as a quadratic form in β (Ruppert, Wand, and Carroll 2003). This allows us to write the penalty in matrix form as

$$\int_{x_1}^{x_n} f''(x)^2 dx = \beta' \mathbf{D} \beta \qquad (3.19)$$

where $\mathbf{D}$ is a matrix of the following form:

$$\mathbf{D} = \begin{bmatrix} 0_{2\times 2} & 0_{2\times k} \\ 0_{k\times 2} & \mathbf{I}_{k\times k} \end{bmatrix}. \qquad (3.20)$$

.

In the above matrix k denotes the number of knots. With a matrix form for the penalty, we can write the penalized spline regression model in matrix form

$$SS(f, \lambda) = ||\mathbf{y} - \mathbf{X}\beta||^2 + \lambda \beta' \mathbf{D} \beta. \qquad (3.21)$$

The hat matrix for the penalized spline has a form where the standard linear regression matrix is altered to accomodate the penalty term. Ruppert, Wand, and Caroll (2003) derive the following hat or smoother matrix for penlized splines:

$$\mathbf{S}_\lambda = \mathbf{X}(\mathbf{X}'\mathbf{X} + \lambda^{2p}\mathbf{D})^{-1}\mathbf{X}' \qquad (3.22)$$

where p is the order of the polynominal of the basis functions. In this form, we see that the penalty term is a scalar λ^{2p} multiplied by the matrix operator $\mathbf{D}$. The trace of $\mathbf{S}_\lambda$, just as in linear regression, represents the degrees of freedom in the spline model and is nearly equivalent to the number of parameters in the spline fit. Due to shrinkage from the penalty term, the degrees of freedom for a penalized spline model will not be an integer. At this point, it should be clear how penalized splines are estimated. The user selects values for λ and p and constructs a model matrix with a set of basis functions identical to that used for a standard spline model. Using these values, we form the matrix $\mathbf{S}_\lambda$ and apply it to the outcome vector to form a set of predictions

$$\hat{\mathbf{y}} = \mathbf{S}_\lambda \mathbf{y}. \qquad (3.23)$$

The predictions from the above equation are the penalized spline estimate that we can then plot. The simplicity of the smoothing spline matrix notation makes programming smoothing splines only slightly more complicated than a cubic spline. The process is the same as for a regression spline model in that y is regressed on a constructed model matrix using least squares. However, while estimating the smoothing spline is feasible with least squares, this method is often not numerically stable. Orthogonal matrix methods, such as either the Cholesky decomposition or a spectral decomposition, are required for numerical stability. An example of basic smoothing spline code is available on the book website.

Since the degrees of freedom are a mathematical transformation of λ, they provide control over the amount of smoothing applied to the data. By selecting the degrees of freedom, the analyst chooses the number of effective local parameters used in the spline estimate. Controlling the penalty using the the degrees of freedom is also more practical instead of selecting a value for λ, since often large differences in λ will translate into fairly small differences in the degrees of freedom. One important difference between penalized splines and standard splines is that the number of knots used now exerts very little influence over how smooth the fit is as the value of λ now controls how smooth the fit is. These penalized splines are commonly referred to as 'smoothing splines'. In the text, we refer to these splines interchangeably as either penalized splines or smoothing splines. At this point, an illustration with smoothing splines is useful to show how they differ from standard splines.

We return to the Congressional elections example to better understand the operation of smoothing splines. Thus far, we have consistently seen a nonlinear relationship between challenger's vote share and support for Perot in the 1992 election. We estimate the relationship between these two variables using smoothing splines to demonstrate how selecting different values for the degrees of freedom affects the fit.

Figure 3.6 contains four smoothing spline estimates where we set the degrees of freedom to 2, 4, 8, and 12 using a cubic spline basis. We hold the number of knots constant across all four fits. For smoothing splines, a large number of knots are placed evenly throughout the range of x such that 10 to 20 x values fall between each knot placement. There are also some simple selection rules to ensure an optimal number of knots. See Ruppert, Wand, and Carroll (2003) for details on knot selection algorithms. Different methods are implemented with different smoothing spline software, and most allow the user to control the number of knots directly as well. The software used in this example simply places knots so that there are 10 to 20 x observations between each set of knots. This implies that for the estimates in Figure 3.6 the models are fit with 31 knots. Given that nine knots produced noticeable overfitting with cubic B-splines, we might expect extreme

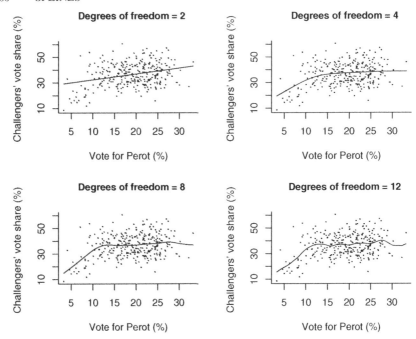

Figure 3.6   Smoothing spline fit to challenger vote share data with four different degrees of freedom values.

overfitting in this illustration. As the reader can see, however, this is not the case. The fit with 2 degrees of freedom is identical to a linear regression fit since there are only two effective parameters: one for the slope and one for the intercept. In the next model, with 4 degrees of freedom, we see the same pattern of nonlinearity found with other spline fits. There are only minor differences between the fit with 4 and 8 degrees of freedom. For the fit with 12 degrees of freedom, we see that considerable variability is caused by using too many parameters, and we overfit the data.

We next demonstrate that the number of knots used in the spline model matters little with smoothing splines. We fit two smoothing spline models to the Congressional election data. In both models, we hold the degrees of freedom constant at 4, but for one model we use four knots, and for the second model, we use 16 knots. The resulting fits are in Figure 3.7. For a cubic spline model, nine knots caused noticeable overfitting, but with the smoothing spline models, however, the results are invariant to the number of knots used. There is very little difference between four and 16 knots for the smoothing spline estimates. Despite

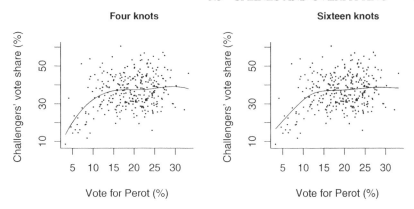

Figure 3.7   Smoothing spline fit to challenger vote share data with constant degrees of freedom and different knot numbers.

the additional parameters that are used for the model with 16 knots, the shrinkage imposed by the penalty term controls the amount of smoothing.

With smoothing splines, we have shifted control over the amount of smoothing from the knots to the tuning parameter, λ, or equivalently the degrees of freedom. The advantage is that the smoothing spline model will provide the same fit with fewer effective parameters, thus reducing the likelihood of overfitting regardless of the amount of smoothing imposed by the analyst. As such, while smoothing splines cannot eliminate the possibility of overfitting, smoothing splines reduce that possibility. Moreover, as we will see in the next section, smoothing splines provide an optimal tradeoff between fit and the number of parameters used. One might also ask: is there a formal method for selection of λ? In fact, there are a number of different methods for choosing λ, which we consider in the next chapter.

3.3.2   Splines as Mixed Models

Random effects are a common solution to unobserved heterogeneity in statistical models. Panel data models, for example, are often susceptible to such heterogeneity. In panel data models, the model intercept varies across the units in the data. Random effects can be used to model the variation across the unit intercepts. In the random effects model, a random shock, drawn from a normal distribution with mean 0 and standard deviation σ^2, is added to the intercept for each observation. This random shock is then integrated out of the likelihood, which removes the heterogeneity.

In statistics, models that include random effects are often referred to as *mixed models* (Pinero and Bates 2000). In the social sciences, a special case of mixed models called multilevel or heirarchical models are used with some frequency. We can use mixed models to estimate splines. In fact, smoothing splines are exactly represented as the optimal predictor in a mixed model framework (Ruppert, Wand, and Carroll 2003). While the mixed model framework has little effect on how we use splines in applied data analysis, the mixed model framework for splines provides two insights. First, it provides an analytic framework for understanding why smoothing splines are the optimal smoother. Second, the mixed model framework represents nonlinearity as unobserved heterogeneity, which helps to clarify the need for nonparametric regression techniques. Next, we (very briefly) outline the mixed model framework and demonstrate how to represent smoothing splines as mixed models. For a full account of splines as mixed models see Ruppert, Wand, and Carroll (2003). For readers unfamiliar with mixed models, this section can be skipped and little will be lost in terms of the practical usage of splines.

In the mixed model framework, we write a linear regression model for Y as:

$$Y_{ij} = \beta_{0j} + \beta_{j1} X_{ij} + e_{ij}. \tag{3.24}$$

For a mixed model, typically each term has two subscripts (more are possible). In this notation, we have i observations nested in j units. This might be i students nested in j schools or i voters nested in j counties. We believe there is heterogeneity across the j units; that is the intercepts and slopes may vary across the j units. If we suspect such variation across the slopes and intercepts, a random set of draws from a normal distribution is added to the intercept and slope parameters in the following manner:

$$\beta_{0j} = \gamma_{00} + u_{0j}$$
$$\beta_{1j} = \gamma_{01} + u_{1j} \tag{3.25}$$

where u_{0j} and $u_{1j} \sim N(0, \sigma^2)$ are the random effects, γ_{00} is the mean outcome across the j units, and γ_{01} is the average slope across the j units. Using either MLE or restricted MLE, we integrate over the random effects to remove their effect from the log-likelihood.

Importantly, a mixed model represents a penalized approach to grouped data as it is a compromise between a parsimonious model and one with too many parameters. With data clustered in units, the analyst can estimate three different statistical models. In the first model, the analyst ignores the clustering within the j units and estimates a regression with the data pooled. The estimates from this 'pooled' model will be biased if the j units differ much, but with the pooled data, there will be little variability in the fit as it pools all the observations. The model estimated with pooled data is the most parsimonious in terms of the number of parameters estimated. At the other extreme, the analyst could estimate one regression model for each of the j units. This option produces too many parameters, and the estimates of the β parameters will be highly variable if there are not very many observations within each j cluster. However, these estimates will be unbiased. The mixed model approach represents a middle ground between these two extremes. The mixed model estimate of the β's is a compromise between ignoring the structure of the data and fully taking it into account by estimating j different models. The mixed model estimate shrinks the estimate from the pooled estimate toward the individual j estimates. The resulting estimates from the mixed model are weighted averages that 'borrow strength' across the units. This shrinkage estimate is superior in mean square error terms to both the pooled and unpooled alternatives. It can be shown that the mixed model estimates are the Best Linear Unbiased Predictors (BLUPs) (Pinero and Bates 2000; Ruppert, Wand, and Carroll 2003). Therefore, if there are significant differences across the j units, the mixed model provides the best fit according to the mean squared error criterion.

Nonlinearity is a similar form of heterogeneity across groups. The data within a set of knots forms each group. A linear fit to the data is identical to a pooled model that ignores any local variation that might exist. This pooled model ignores the underlying structure and uses a single parameter to summarize the relationship. This fit is the most parsimonious but ignores difference in the fit between x and y across the groups of data between knots. If the variation is substantial the linear fit is biased. A standard spline model with knots for each x value represents model with no pooling. Here, we use a parameter for each x and y value. Such a model produces an estimate that is very faithful to the local variation, but the estimates can be highly variable if there is too little data between each set of knots. The smoothing spline has the same structure as a mixed model as it takes the local variation into account but also borrows strength across all the knot segments. This implies that the smoothing spline should have the best fit according to the mean squared error criterion. The smoothing spline estimate is the BLUP for a model of heterogeneity due to nonlinearity as it shrinks the global estimate toward a model with highly local fits.

The regression spline model (with a linear basis) for f is:

$$f(x_i) = \beta_0 + \beta_1 x_i + \sum_{k=1}^{K} \beta_k^*(x_i - c_k)_+ + \varepsilon \qquad (3.26)$$

where $(x_i - c_k)_+$:

$$(x_i - c_k)_+ = \begin{cases} 0 & x \le c_k \\ x - c_k & x > c_k \end{cases} \tag{3.27}$$

represents a piecewise linear fit with knots at c_k. Now, rewrite each term in Equation (3.26) in matrix form:

$$\mathbf{X} = \begin{bmatrix} 1 & x_1 \\ \vdots & \vdots \\ 1 & x_n \end{bmatrix} \tag{3.28}$$

$$\mathbf{Z} = \begin{bmatrix} (x_1 - c_1)_+ & \cdots & (x_1 - c_k)_+ \\ \vdots & \ddots & \vdots \\ (x_n - c_1)_+ & \cdots & (x_n - c_k)_+ \end{bmatrix} \tag{3.29}$$

$$\boldsymbol{\beta} = \begin{bmatrix} \beta_0 \\ \beta_1 \end{bmatrix} \tag{3.30}$$

$$\boldsymbol{\beta^*} = \begin{bmatrix} \beta_1^* \\ \vdots \\ \beta_k^* \end{bmatrix} \tag{3.31}$$

and

$$\boldsymbol{\varepsilon} = \begin{bmatrix} \varepsilon_1 \\ \vdots \\ \varepsilon_n \end{bmatrix}. \tag{3.32}$$

The vector, $\boldsymbol{\beta^*}$, represents the coefficients that define the piecewise functions. We combine these terms to write Equation (3.26) in matrix form:

$$\mathbf{y} = \mathbf{X}\boldsymbol{\beta} + \mathbf{Z}\boldsymbol{\beta^*} + \boldsymbol{\varepsilon}. \tag{3.33}$$

To write the linear spline model as a mixed model requires:

$$\mathbf{u} = \begin{bmatrix} u_1 \\ \vdots \\ u_K \end{bmatrix}, \tag{3.34}$$

a vector of random effects, with each element drawn from $N(0, \sigma^2_{u_k})$. We rewrite Equation (3.33) as the following mixed model:

$$\mathbf{y} = \mathbf{X}\boldsymbol{\beta} + \mathbf{Z}\mathbf{u} + \boldsymbol{\varepsilon}. \qquad (3.35)$$

The difference in the models is that a random effect is placed on the location of each knot. The solution to Equation (3.35) is:

$$\hat{\mathbf{f}} = \mathbf{C}(\mathbf{C}'\mathbf{C} + \lambda^2 \mathbf{D})^{-1}\mathbf{C}'\mathbf{y} \qquad (3.36)$$

where $\mathbf{C} = [\mathbf{X}\ \mathbf{Z}]$, $\mathbf{D} = \text{diag}(0,0,1,\ldots,1)$, and $\lambda^2 = \sigma^2_\varepsilon / \sigma^2_u$. Ruppert, Wand, and Carroll (2003) show that Equation (3.35) is equivalent to

$$\text{SS}(f, \lambda) = \sum_{i=1}^{n} y_i - f(x_i)^2 + \lambda \int_{x_1}^{x_n} f''(x)^2 dx \qquad (3.37)$$

which is the smoothing spline representation.

What are the implications of this equivalence between smoothing splines and mixed models? First, this implies that the smoothing spline fit is the BLUP for the nonlinear estimate between x and y. As such, the smoothing spline fit should have the lowest mean squared error if nonlinearity is present. Therfore, the smoothing spline provides the best tradeoff between fit and the number of parameters used. Second, the mixed model framework allows us to conceptualize the nonlinearity between x and y as unobserved heterogeneity. If such unobserved heterogeneity is present and goes unmodeled, specification error results. Non-parametric regression provides us with a tool to model this heterogeneity. The logic behind nonparametric regression then is no different from that behind mixed models. Nonlinearity is a form of heterogeneity, and for such heterogeneity, the smoothing spline will provide the best estimate that balances fit with the number of parameters used. The mixed model representation of smoothing splines also allows smoothing splines to be easily translated into a Bayesian estimation framework, and it allows for the incorporation of smoothing splines into standard mixed models. We provide examples of both in Chapter 7.

3.3.3   Final Notes on Smoothing Splines

In general, analytic proofs for the properties of nonparametric regression are unavailable. Wood (2006), however, proves that cubic smoothing splines provide the fit with the least amount of error. The full proof is not overly complex but is lengthy, so we omit it here and refer the reader to Green and Silverman (1994) and Wood (2006). Below, we provide a sketch of the proof. The goal for any nonparametric regression model is to select the estimate of f that minimizes the squared error between the fit and the data. We must select an estimate of f that is continuous on $[x_1, x_n]$ and has continuous first derivatives and is smoothest in the sense of minimizing

$$\int_{x_1}^{x_n} f''(x)^2 dx.$$

If $f(x)$ is the smoothing spline fit, it is the function that minimizes:

$$\sum_{i=1}^{n} y - f(x)^2 + \lambda \int_{x_1}^{x_n} f''(x)^2 dx. \tag{3.38}$$

Why? In short, the interpolant properties of smoothing splines ensure that no other function will have a lower integrated squared second derivative. Smoothing splines have the best properties in that no other smoothing function will have a lower mean squared error. Therefore, on analytic grounds, smoothing splines are to be preferred to other smoothers.

Finally, it is possible to use splines to smooth in more than two dimensions. The same caveats apply. The curse of dimensionality remains a problem, and it remains impossible to interpret fits in more than two dimensions. In general, however, most work in statistics uses splines to smooth in two dimensions rather than local polynomial regression.

3.3.4 Thin Plate Splines

We outline one last type of spline model. This form of spline, thin plate splines, is generally unnecessary for estimating simple nonparametric models or even most semiparametric models. Thin plate splines are primarily used for smoothing in more than one dimension, but they are also useful for estimating spline models with Markov Chain Monte Carlo (MCMC) methods.

With the cubic smoothing spline models considered thus far, the smoothing penalty is placed on the second derivative of the smoothed fit $f(x)$. It is possible to penalize any derivative of the smoothed fit subject to constraints imposed by the dimension of x. Thus, thin plate splines are a more general version of smoothing splines where the penalty placed in the spline fit can be placed on any order of derivative. The formal representation of a thin plate spline is as follows:

$$SS(f, \lambda) = \sum_{i=1}^{n} y_i - f(x_i)^2 + \lambda \int_{x_i}^{x_n} f^m(x_i)^2 dx. \tag{3.39}$$

Notice that the penalty on the derivative is now generalized. Here, m is the mth derivative such that $2m > d$, where d is the dimension of x_i. This implies that moving beyond the second derivative requires smoothing in more than one dimension. The full representation of penalties for thin plate spline is fairly complex, but see Wood (2006) for an introduction. Thin plate splines come at a heavy computational cost as they have as many unknown parameters as there are data points. Typically some low rank approximation is used to avoid this problem. Unfortunately, even low rank thin plate splines are difficult to estimate with data sets over 1000 cases. While there are software packages that implement low rank thin plate regression splines, they offer few advantages for most estimation problems. However, low rank thin plate splines are useful when used with MCMC methods. Thin plate splines also work well when smoothing in more than one dimension.

3.4   Inference for Splines

Inference for splines is identical to inference for local polynomial regression smoothing. For confidence bands, we calculate pointwise standard errors from $\mathbf{S}$, the smoothing matrix, and plug these standard errors into the usual 95% confidence interval formula and plot the result along with the spline fit. We can also test spline models against linear models with or without transformations using F-tests. In this section, we develop the details for inference with splines and provide two illustrations.

Confidence Bands

For splines models, how we define $\mathbf{S}$ depends on whether one is using a smoothing spline or not. Recall that cubic and natural spline models are simply regression models with an expanded model matrix. In the last chapter we defined $\mathbf{S}$ as similar to the $\mathbf{H}$ or hat matrix from a regression model. Since cubic and spline models *are* regression models, then $\mathbf{S}$ is equivalent to the hat matrix: $\mathbf{X}(\mathbf{X}'\mathbf{X})^{-1}\mathbf{X}'$. The only difference is that we have altered the model matrix $\mathbf{X}$ using a set of basis functions. For smoothing splines, $\mathbf{S}$ is more complex. We defined the $\mathbf{S}$ matrix for smoothing splines as $\mathbf{X}(\mathbf{X}'\mathbf{X} + \lambda^{2p}\mathbf{D})^{-1}\mathbf{X}'$. To calculate confidence bands, either form of the $\mathbf{S}$ matrix is used in an identical manner. In the discussion that follows, we assume that λ is fixed and $\mathbf{S}$ represents either formulation of the smoothing matrix. Under these assumptions, the covariance matrix for the fitted vector $\hat{\mathbf{f}} = \mathbf{S}\mathbf{y}$ is

$$\text{cov}(\hat{\mathbf{f}}) = \mathbf{S}\mathbf{S}'\sigma^2. \tag{3.40}$$

If we have an unbiased estimate of σ^2 and the sample size is large enough, we form pointwise confidence interval bands using ± 2 times the square root of the diagonal elements of $\mathbf{S}\mathbf{S}'\sigma^2$. We still need an estimate for σ^2, and we use the same method we used with local polynomial regression. If the estimate of f is unbiased, an unbiased estimator of σ^2 is

$$\text{RSS}/df_{\text{res}} \tag{3.41}$$

where RSS is the residual sum of squares which is defined as: $\sum_{i=1}^{n}[y_i - \hat{f}(x_i)]^2$ and df_{res} is the residual degrees of freedom:

$$df_{\text{res}} = n - 2\text{tr}[\mathbf{S} + \text{tr}(\mathbf{SS'})]. \tag{3.42}$$

In practice, these pointwise bands provide reasonable estimates of variability for the spline fit. In terms of statistical theory, however, these variability estimates have some limitations. First, it would be desirable to account for variability in $\hat{\sigma}$ as an estimate of σ, though this variability should be small so long as the sample size is large. If n is small, one should replace the Normal critical value of 2 with a t-distribution critical value with k degrees of freedom, where k is the number of parameters in the model (the trace of $\mathbf{S}$). More importantly, we must assume that the estimate of $f(x)$ is approximately unbiased. This assumption is difficult to test, and given what we know about nonparametric regression, we must assume that there is some bias in the estimate of f. The nonparametric fit is a tradeoff between bias and variance, and a fit with very little bias will be highly variable, so we must accept some bias to reduce the variance. Since we never observe the true smooth function, it is difficult to judge the level of bias in the fit. For these pointwise variabilty bands, it is often not unreasonable to assume the fit is approximately unbiased. If so, these variability bands can be interpreted as confidence bands. It would be better, however, if we could account for the possible bias in the estimate of f in the estimation of the variability bands. Again, smoothing splines are superior in this respect as they allow for bias adjustment to the confidence bands. There are broadly two different methods for the estimation of bias adjusted variability bands for splines. One is based on Bayesian principles, and the other relies on the mixed model framework. We, first, discuss the Bayesian variability bands.

Splines can be derived from a Bayesian smoothing model that assumes f is a sum of underlying random linear functions and an integrated Weiner process. Wahba (1983) and Nychka (1988) provide full details on the Bayesian derivation of splines. Importantly, they demonstrate that the variability bands from the Bayesian model take into account possible bias in the estimate of f. Using $\hat{\sigma}\mathbf{S}_\lambda$ to estimate the variance for f produces bias adjusted variability bands that are equivalent to the Bayesian confidence bands (Hastie and Tibshirani 1990). Not all smoothing software estimates bias corrected variability bands in this fashion. For example, Wood (2006) suggests several refinements and implements them in his software for smoothing. In general, these refinements alter the estimates of the variability bands little.

When splines are estimated in a mixed model framework, the bias adjusted variance estimate, $\hat{\sigma}^2_{adj}$ is

$$\hat{\sigma}^2_{adj} = \hat{\sigma}_\varepsilon \sqrt{\mathbf{C}_x \left(\mathbf{C}'\mathbf{C} + \frac{\sigma^2_\varepsilon}{\sigma^2_u} \mathbf{D} \right)^{-1} \mathbf{C}'_x} \qquad (3.43)$$

where $\mathbf{C} = [\mathbf{X}\ \mathbf{Z}]$, $\mathbf{D} = \mathrm{diag}(0,0,1,\dots,1)$. In the mixed model framework, the predictions are conditioned on the random effects to account for possible bias in the estimate of f (Ruppert, Wand, and Carroll 2003). The mixed model variability bands typically differ little from those estimated under the Bayesian framework.

We return to our example of challenger's vote share and support for Perot to illustrate the estimation of confidence bands for spline models. First, we estimate the spline fit between the two variables using a natural cubic spline fit and then estimate pairwise standard errors and calculate 95% confidence bands for the spline fit. The result is in Figure 3.8. The estimated confidence bands indicate that the spline estimate is fairly precise. The confidence bands here differ little from those for local polynomial regression. Importantly, these confidence bands are not bias adjusted.

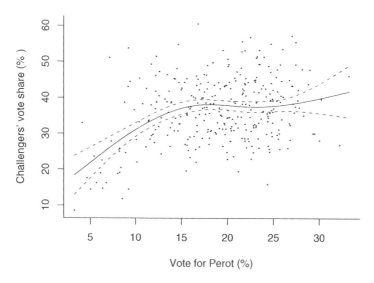

Figure 3.8 Confidence bands for natural cubic spline fit to challenger's vote share.

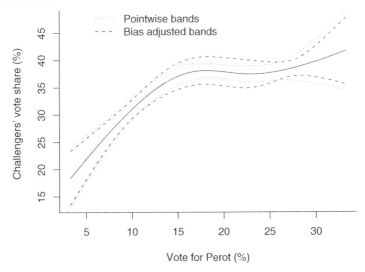

Figure 3.9   A comparison of pointwise variability bands and bias adjusted confidence intervals.

In Figure 3.9, we plot the Bayesian bias adjusted confidence bands along with the standard pointwise variability bands for the fit between challengers' vote share and support for Perot. The difference is negligible. The chief difference is that the bias adjusted bands provide a confidence interval for the fit while the variability bands are a pointwise estimate.

Hypothesis Tests

We can also perform hypothesis tests. Since the spline model nests models with global fits, we can rely on approximate F-tests for hypothesis tests (Hastie and Tibshirani 1990). Again to test for a statistically significant effect, we test the spline model against a model with only a constant. As before, we often test the spline model against a model with a global linear fit or against models with power transformations on the right hand side. If RSS_1 and RSS_2 are the residual sum of squares from a restricted model and the spline model respectively, an approximate F-test is:

$$\frac{(RSS_1 - RSS_2)/(df_{res2} - df_{res1})}{RSS_2/(n - df_{res2})} \sim F_{df_{res2}-df_{res1}, n-df_{res2}}. \qquad (3.44)$$

This is most useful as a means of testing for nonlinearity and assessing the adequacy of power transformations. If there is no difference between a spline fit and a simpler model, the simpler model is preferred for reasons of parsimony. Consequently, there is no need to use nonparametric methods unless the evidence from the data suggests that they are necessary.[4]

Next, we test various hypotheses about the nature of the relationship between the challengers' vote share and support for Perot. We compare the smoothing spline model to a model with only a constant to test whether the effect of support for Perot is significantly different from zero. Not surprisingly, we find that the relationship is highly significant as the test statistic is 17.14 on 3 and 311 degrees of freedom.[5] We also test the spline model against a global linear fit, and the F-test test statistic of 10.31 on 2 and 310 degrees of freedom is highly statistically significant ($p < 0.001$). The results from this test indicate that the relationship between challengers' vote share and support for Perot is sufficiently nonlinear that a global linear model is inadequate. Finally, we test the spline fit against quadratic and logarithmic transformations of the support for Perot variable. In both instances, the F-test indicates that the spline fit is superior. As a result, a spline fit is recommended over these transformations, which do not adequately model the nonlinearity.

Derivative Plots

While nonparametric fits do not allow for the same form of concrete interpretation as parametric models, there are other methods of interpretation available. For example, we can plot the first derivative for the nonparametric estimate. The first derivative, in general, tells us about the rate of change between x and y, and we can calculate the first derivative along the entire range of the nonparametric fit and plot the result to further understand the dependency between the two variables. For example, a plot of the first derivatives provides insight into the following questions for the Congressional elections example:

- How fast does challengers' vote share increase?

- When does the effect of Perot support start to level off?

- Does the effect of support for Perot level off or does it have a significant negative effect on challengers' votes share for higher values?

---

[4]The F-test is only considered to be approximate for smoothing splines. However, as Hastie and Tibshirani (1990) note general investigation indicates that the approximation is fairly good. Bootstrapping techniques can be used to replace the parametric F-test assumption. See Chapter 8 for details on bootstrapping nonparametric models.

[5]I set the smoothing parameter, λ, to 4 and used a cubic basis function. This was based on earlier analyses that indicated that a fit with 4 degrees of freedom provided a reasonable fit.

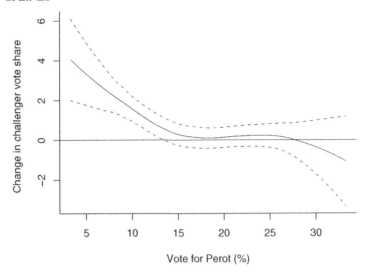

Figure 3.10   Plot of first derivative for the effect of support for Perot on the challenger's vote share.

Figure 3.10 contains a plot of the first derivative for the spline estimate. In the plot, challengers gain vote share quite rapidly until support for Perot reaches approximately 15%. After that, there is no significant change in the vote share for challengers until support for Perot exceeded 25% in the district. However, at this point the confidence bands are so wide, one cannot place much emphasis on the deviation from 0. This is clearly due to the fact that there were very few Congressional districts where support for Perot exceeded 25%. Derivative plots are useful since they often allow for a richer interpretation of nonparametric regression models.

3.5   Comparisons and Conclusions

Thus far, we have presented the reader with several different nonparametric regression models. In this section, we provide a brief comparison of these models. It is logical to ask whether one form of nonparametric regression is preferable to another. For many applied problems, there aren't strong reasons to prefer one smoother over another, but the smoothing spline does have some advantages. It is the only smoother that places a penalty on the number of parameters used in the fit. While the critique of nonparametric regression as prone to overfitting is perhaps overstated, the smoothing spline is designed to meet this criticism directly.

Moreover, the mixed model representation of smoothing splines provides an analytical foundation for these smoothers.

We present some basic simulation evidence to demonstrate how well different smoothers approximate a highly nonlinear functional form. For the simulation, we generated a function for f and fit four different smoothers, two local polynominal smoothers and two spline smoothers, to the simulated data. The simulated functional form is

$$y = \sin^3(2\pi x^2) + \varepsilon. \tag{3.45}$$

The result will be a cyclical pattern between x and y. Such functional forms are rare in the social sciences but make difficult estimation problems for nonparametric regression models. The four nonparametric fits are plotted in Figure 3.11.

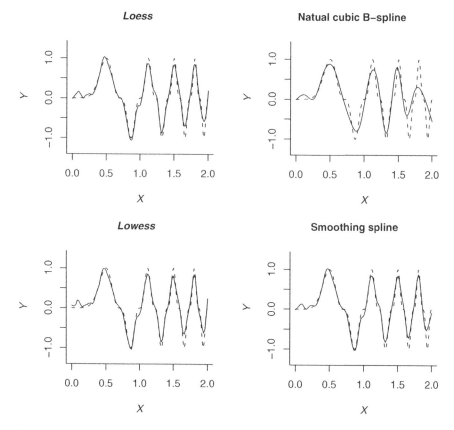

Figure 3.11   Comparison of smoother fits. Dotted line is true functional form, solid line is nonparametric estimate.

The first nonparametric model is the *loess* smoother. Using the visual trial and error method, we selected a value of 0.1 for the span. The *loess* fit is good though some minor undersmoothing occurs as it misses some of the peaks and valleys. In the upper right hand panel is a natural cubic B-spline with the knots chosen by AIC values. This model displays noticeable undersmoothing and provides a poor estimate for the upper range of x. In the lower left panel is a *lowess* estimate. The *lowess* estimate is almost identical to that from *loess*, so again we see that the additional weights used in *lowess* add little to the estimate. Finally, we estimate a smoothing spline using 31 degrees of freedom also selected through visual trial and error. The smoothing spline closely matches the nonlinearity and has only minor undersmoothing. The performance for the smoothing spline is quite similar to both the LPR smoothers. The simulation evidence suggests that both the LPR models and the smoothing spline are good choices. For basic smoothing of scatterplots, there is little reason to prefer one smoother over another. Only in rare instances should we find that different smoothers provide different estimates. As we will see in the next chapter, however, there are additional reasons to prefer the smoothing splines. Moreover, for semiparametric regression models, smoothing splines are preferable.

Smoothers are, by themselves, little more than powerful diagnostic tools. It is hard to imagine conducting an analysis with nonparametric regression alone. Smoothers, however, are particularly useful for deciding whether the relationship between two variables is linear or not. A visual examination of a scatterplot without the use of a smoother is a poor second option when smoothers allow you to easily see whether a relationship is linear or not. As we will see, smoothers are more powerful when used in conjunction with standard parametric models.

3.6   Exercises

Data sets for exercises may be found at the following website: http://www.wiley.com/go/keele_semiparametric.

1. The data in demo2.dta describe political demonstrations across 76 nations. The dependent variable is the percentage of activity in the country that can be classified as political demonstrations. The data set also contains two independent variables. One which records the level of unionization in the nation and a second which records the level of inflation. (NB: Don't name your data frame demo, R will confuse the demo variable with the data frame.)

   (a) First, use cubic splines to study the relationship between the demonstrations and unionization. Compare the cubic spline model to a natural cubic spline model. Do you see in differences in the fit near the boundaries of the data?

(b) Using natural cubic splines experiment with both the number of knots and the placement of the knots. How many knots are required before the fit appears to be clearly overfit? Select the number of knots using AIC.

(c) Compare a spline fit to a *lowess* fit. Are the results any different? Next, compare a smoothing spline model to a natural cubic spline model.

(d) Calculate 95% confidence interval bands and plot the relationship with CI bands.

(e) Test whether the relationship is statistically significant and whether there is a significant amount of nonlinearity. Does the spline model improve upon linear fits with quadratic or logarithmic transformations?

(f) Add inflation to the model and plot the joint nonparametric effect. Does inflation improve the fit of the model?

(g) Finally, develop computer code for your own spline smoother for this data. Start with piecewise linear functions. Can you roughly approximate the spline models with piecewise linear functions? Try piecewise quadratic fits.

2. The `forest-sub.dta` data on evironmental degradation across countries. The data set contains a measure of deforestation, a measure of democracy (-10 to 10) and a measure of GDP per capita. Scholars in international relations often debate whether the effect of this democracy scale is constant. As a result, some analysts recode the scale to dummy variables for democracy (6 to 10) and autocracy (-6 to -10). Investigate whether splines can capture the effect of democracy on deforestation. First, use a scatterplot to examine the relationship between democracy and deforestation. Then fit a spline model to the data. Add confidence bands to the plot. Test this fit against a linear, quadratic and logarithmic models. Which model do you prefer? Does the spline model reveal anything substantive about the effect of democracy? What does it imply if the effect is linear as opposed to nonlinear?

4

Automated Smoothing Techniques

Regardless of which smoother used, the analyst must choose the amount of smoothing to apply to the data. When using a local polynomial regression smoother, the analyst must decide how to set the span, and with splines, the analyst must choose either the number of knots or the degrees of freedom. The basic method of smoothing selection is one of visual trial and error. While this method works well in most instances, smoothers are often criticized as being little more than guesswork since they require the analyst to make a choice that can induce either bias or inflate the variance of the fit. Statisticians, however, have developed a number of automated smoothing techniques that obviate the need to use visual smoothing selection methods. With automated smoothing, we estimate the amount of smoothing to be applied (the span or a value for λ) from the data and use this estimate to fit the nonparametric regression model. With automated smoothing, the analyst is not required to make any decisions about the smoother estimation process. We argue that in many contexts automated smoothing techniques are a good choice for applied analysts. This chapter starts with an outline of how these automated smoothing techniques work. We begin with a discussion of automated smoothing techniques for local polynomial regression and splines. The chapter concludes with a number of empirical examples to demonstrate the use of automated smoothing.

Semiparametric Regression for the Social Sciences   Luke Keele
© 2008 John Wiley & Sons, Ltd

4.1   Span by Cross-Validation

For local polynomial smoothers, we would like to estimate a span parameter that minimizes both bias and variance in the smoothed fit. The span parameter, however, is an external parameter which implies that we must compare a series of nonnested models each with a different span setting. This suggests the use of a model selection methods such as cross-validation. Cross-validation is a general technique for assessing model fit based on resampling that can be applied to most statistical models. Cross-validation relies on the residual sum of squares (RSS) as a measure of model fit.[1] While the RSS is a measure of predictive ability, it cannot be used for model selection since that would imply that values of y are used to predict y, and as a result, the RSS will always select the model which uses the most degrees of freedom. With cross-validation, the analyst partitions the data to avoid y being predicted by its own values. In the simplest form of cross-validation, the data are randomly split in half to form two separate data sets. In the first data set, called the training data, the analyst is allowed to select a model specification by any means (stepwise methods, dropping terms with low p-values, AIC, etc.). The model specification derived from the training data is then applied to the second data set to see how well it predicts in the new sample. There are a variety of other ways to partition the data for cross-validation though most of these methods require fairly large sample sizes. Interested readers should consult Efron and Tibshirani (1993). Leave-one-out cross-validation is probably the most commonly used method to partition the data since it works well with most any sample size. With leave-one-out cross-validation, one observation is randomly selected and then omitted from the data set. The analyst then fits one of the possible models to this slightly truncated data set and calculates a measure of fit. Next, a new data point is dropped, and the measure of fit is calculated again. This process is repeated as each of the data points is removed from the data set. The cross-validation score is the averaged measure of model fit and can be used to compare different model specifications. While one could use different data partitions, leave-one-out cross-validation is most often used to select a span parameter for local polynomial regression.

To select the span by cross-validation, we omit the ith observation from the data and fit a local regression model with the span set at s. We denote this estimate as $\hat{f}_s(x_{-i})$ and we compute the squared difference between this nonparametric estimate and y. This process is repeated for all N observations. We compute the

---

[1] The RSS for a linear model is $\sum e_i^2$, where e_i is the residual term.

cross-validation score by averaging the N squared differences

$$\mathbf{CV}(s) = \frac{1}{n} \sum_{i=1}^{n} [y_i - \hat{f}_s(x_{-i})]^2 \tag{4.1}$$

where $\hat{f}_s$ is the nonparametric estimate of y for span s minus one observation. The value of s that minimizes this function is considered to be the optimal amount of smoothing to apply to the local regression fit. Leave-one-out cross-validation is relatively computationally intensive since we have to estimate n nonparametric fits for each value of s. For a data set with 100 observations and 10 values for s, that implies estimating a *lowess* model one thousand times. For larger data sets, using cross-validation becomes cumbersome. Instead, we can use an approximation to cross-validation called generalized cross-validation (GCV). Craven and Wahba (1979) developed the GCV score as a less computationally-intensive alternative to cross-validation. The generalized cross-validation (GCV) score is

$$\mathbf{GCV}(s) = \frac{\sum_{i=1}^{n} [y_i - \hat{f}_s(x_i)]^2}{(n - df)^2} \tag{4.2}$$

where df is the fitted degrees of freedom from the nonparametric model. This GCV score is computed for several different values of s, and the value with the lowest GCV score is selected. The GCV is essentially an approximation for cross-validation, where we compare the smoothed fit to the observed y while adjusting for the degrees of freedom used in the nonparametric estimate.

Unfortunately, cross-validation in either form tends to perform poorly when used with local polynomial regression. The GCV score is an estimate and as such is subject to sampling variability, and even in moderate sample sizes, the estimation uncertainty can be quite large. In general, GCV tends to underestimate s when used with local polynomial regression models unless we have fairly large sample sizes. There are several other bandwidth selection methods that have been designed to improve upon cross-validation and GCV. As Loader (1999) demonstrates, however, these other methods often offer little improvement over cross-validation and have drawbacks of their own. He concludes that GCV remains the most useful tool for selecting the span parameter. With social science data sets, automated routines for local polynomial regression tend to overfit the data. In general, automated smoothing techniques work far better with spline models, which has probably contributed to the increased focus on splines over LPR in statistical research. Finally, whatever its problems, cross-validation routines are unavailable for most software implementations of *loess* and *lowess*.

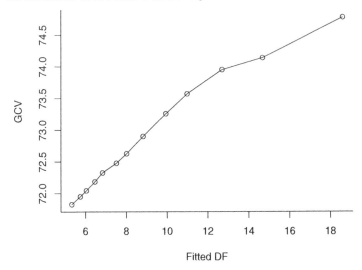

Figure 4.1   GCV plot for local polynomial regression span selection.

We return to the data on the 1992 House elections to illustrate the use of cross-validation techniques with LPR smoothers. Using the visual method of span selection, we found in Chapter 2 that 0.30 provided a reasonable fit, but that increasing the span to either 0.40 or 0.50 changed the estimate little. A value of 0.20, however, appeared to overfit the data. Here, we use GCV to select the span for a *lowess* fit. We select a range of values for s, and we use each value of s to estimate the GCV score. We then plot the GCV scores, and the point at which the GCV scores reach a minimum is the optimal span. Figure 4.1 contains a plot of the GCV scores for values of s from 0.20 to 0.80 in increments of 0.05.

First, note that the x-axis is labeled as the degrees of freedom instead of the span parameter. This is a result of the software defaults used in this example. Each point represents a GCV score for a span parameter increment, and each point on the line represents a span value. For example, the span setting of 0.40 represents a use of approximately six degrees of freedom for the LPR estimate. In the plot, a clear minimum emerges at 0.20. We plot the local likelihood fit between challengers' vote share and support for Perot using a span value of 0.20.[2] The resulting nonparametric estimate is plotted in Figure 4.2. The *lowess* estimate is clearly overfit as it is very wavy and accentuates several idiosyncratic

---

[2]We use a local likelihood fit since automatic span selection is only implemented with local likelihood software. We select a local regression likelihood making the fit equivalent to *lowess*.

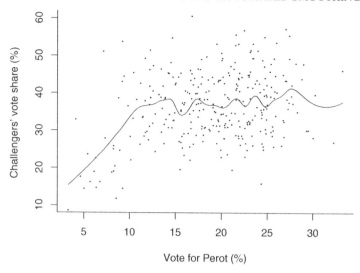

Figure 4.2   *Lowess* fit with span selection via GCV.

aspects of the data. As Loader (1999) notes, GCV estimates with local regression techniques can be data sensitive, working well with some data but not others. In general, analysts are often better off simply using the visual method of selection for LPR smoothers. As we will see, automated smoothing tends to work better with splines.

4.2   Splines and Automated Smoothing

While automated smoothing techniques can be applied to any type of spline, smoothing splines provide a more natural framework for automated smoothing. For both local polynomial regression and standard spline models, the span and number of knots are external parameters that are applied to the model. For smoothing splines, however, the amount of smoothing is a parameter within the model. Therefore, when we ask what is the 'best' value of λ, the smoothing parameter, we can use standard statistical principles of model performance. The optimal value of λ will be one that minimizes the model mean square error. In theory, we could estimate λ as a function of the nonparametric regression mean square error (MSE)

$$L(\lambda) = \frac{1}{n} \sum_{i=1}^{n} [f(x_i) - f_\lambda(x_i)]^2 \tag{4.3}$$

where $f_\lambda(x_i)$ is a smoothing spine fit for a given value of λ. This estimate of λ will depend on the unknown true regression curve and the inherent variability of the smoothing estimator. Estimation of λ through this function would have the smallest MSE and be optimal. Since the true $f(x_i)$ is unobserved, however, we cannot directly estimate λ from Equation (4.3). There are two alternative methods for estimating λ, one using maximum likelihood estimation and the other using cross-validation. Both methods tend to produce reasonable estimates of λ.

4.2.1   Estimating Smoothing Through the Likelihood

In Chapter 3, we saw that the smoothing spline model is equivalent to a mixed model, where random effects are placed at each knot to ensure a reasonably smooth fit. The mixed model estimating equation for the smoothing spline model is

$$\frac{1}{\sigma_e^2}||\mathbf{y} - \mathbf{X}\beta - \mathbf{Z}\mathbf{u}||^2 + \frac{\lambda^2}{\sigma_e^2}||\mathbf{u}||^2. \tag{4.4}$$

One advantage of the mixed model representation is that the smoothing penalty is part of the likelihood for the smoothing spline mixed model. Therefore, λ is simply another parameter to be estimated as part of the likelihood. Ruppert, Wand, and Carroll (2003) show that for a cubic smoothing spline, the estimate of λ is

$$\hat{\lambda} = (\hat{\sigma}_\varepsilon^2/\hat{\sigma}_u^2)^{3/2} \tag{4.5}$$

where $\hat{\sigma}_\varepsilon^2$ and $\hat{\sigma}_u^2$ are the estimated variances of the random effects and the error term of the model. Estimating λ through the likelihood is appealing from a statistical standpoint, since the smoothing parameter becomes another parameter estimated from the data in the likelihood. Smoothing spline models estimated via other means must rely on model selection techniques such as cross-validation.

4.2.2   Smoothing Splines and Cross-Validation

The process for choosing λ through cross-validation is identical to selecting the span with cross-validation. For a given value of λ, one observation is selected at random to be removed from the data set. The smoothing spline model is fit to this slightly truncated data set. The model prediction errors are calculated, and this is repeated as each observation is dropped in turn. The cross-validation score is calculated as the average of the individual model prediction errors. The analyst chooses the value of λ with the smallest cross-validation score. Therefore, the ordinary cross-validation score is

$$\mathbf{CV}(\lambda) = \frac{1}{n} \sum_{i=1}^{n} [y_i - \hat{f}_\lambda(x_{-i})]^2. \tag{4.6}$$

To reiterate, the cross-validation score $\mathbf{CV}(\lambda)$ results from leaving out one datum at a time, fitting the model and calculating the squared difference between this model fit and y_i. These squared differences are then averaged. This process must be repeated for a plausible range of λ values, and we select the value for λ that minimizes the cross-validation score. Ordinary cross-validation, however, remains computationally intensive and for spline models suffers from an invariance problem (Wood 2006; Wahba 1990). That is, the ordinary cross-validation score is not invariant to transformations of the data when used with smoothing splines. Cross-validation may return different optimal λ values for the same models after an orthogonal transformation has been made to either x or y. As result, GCV is used instead. For smoothing spline models, the generalized cross-validation score is

$$\mathbf{GCV}(\lambda) = \frac{\sum_{i=1}^{n} [y_i - \hat{f}_\lambda(x_i)]^2}{[1 - n^{-1} tr(\mathbf{S})]^2} \tag{4.7}$$

where $\mathbf{S}$ is the smoother matrix as defined for smoothing splines in Chapter 3. The term $\hat{f}_\lambda(x_i)$ denotes a smoothing spline estimate fitting all the data for a given value of λ. While GCV is considered superior to ordinary cross-validation for automated smoothing, the differences between the two are often trivial in practice. Both cross-validation methods perform better with smoothing splines than with local regression. The smoothing spline penalty tends to prevent the type of overfitting that we witnessed when GCV was used with local polynomial regression.

Regardless of whether one uses the cross-validation or the likelihood approach, the confidence bands can be estimated to reflect the estimation uncertainty for λ. When λ is estimated via GCV, a Bayesian confidence interval is used. By imposing a penalty on the smoothed fit, we are imposing a prior belief about the characteristics of the model, and Bayesian confidence intervals reflect this prior belief (Wood 2006; Wahba 1983; Silverman 1985). In the likelihood context, the confidence intervals are estimated from the mixed model variance components (Ruppert, Wand, and Carroll 2003). Using either of these methods ensures that estimation uncertainty for λ is reflected in the confidence bands.

The reader might wonder if there is any reason to prefer the likelihood approach over the use of cross-validation. While the likelihood approach is to be preferred in terms of statistical elegance – λ becomes another parameter in the model to be estimated – in practice there is little reason to prefer one method over the other. Ruppert, Wand, and Carroll (2003) conducted Monte Carlo simulations to compare the performance of the two methods and found few differences. Since the performance differs little, the choice between the two methods is reduced to one's software preference. The likelihood approach has one software implementation, and the cross-validation approach has been implemented for several other spline routines. See the software appendix for further details.

We return to the 1992 Congressional elections data to demonstrate the use of automated smoothing techniques with smoothing splines. We again estimated the challenger's vote share as a function of support for Perot, but now we use automated smoothing methods. We fit two smoothing spline models using GCV and mixed model methods for each one as a comparison. Figure 4.3 contains the resulting smoothing spline fits. In the left panel of Figure 4.3 is the fit with the amount of smoothing selected by GCV, and in the right panel is the fit with the amount of smoothing estimated through the likelihood. The GCV degrees of freedom estimate is 3.876, and the mixed model estimate is 3.523 degrees of freedom; since the degrees of freedom settings are estimates, slight differences are to be expected. Visual inspection of the plots, however, reveals no obvious differences across the two spline models. Two points about this illustration are noteworthy. First, neither method appears to overfit the data as both provide reasonably smooth fits. Second, the differences between the likelihood and cross-validation methods are minimal. Again, there is little reason to prefer one method over the other beyond software considerations.

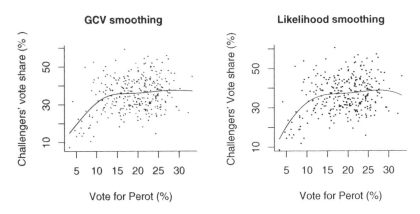

Figure 4.3   Smoothing spline fits with automatic smoothing selection.

4.3   Automated Smoothing in Practice

We now discuss automated smoothing methods relative to choosing the amount of smoothing manually. There is not a single definitive reason to prefer automated smoothing, but automated smoothing is often a useful tool. Here, we consider some of the merits of automated smoothing.

The choice about the level of smoothness to impose introduces a unique form of uncertainty into the nonparametric regression modeling process. This has led some analysts to see the use of smoothers as more akin to art than science. In fact, some authors in political science have described the knot selection process as 'controversial' (Box-Steffensmeier and Jones 2004). This is an unfair criticism of smoothers. In practice, even with visual methods of selection, the nonparametric estimate should be fairly invariant to the amount of smoothing imposed on the fit. It usually requires fairly extreme values for the smoothing parameter to drastically alter the nonparametric regression estimate. Still one advantage of automated smoothing methods is their use nullifies any objections to smoothing being more art than science. When the amount of smoothing is estimated from the data, it eliminates any perception that the analyst is influencing the fit through the smoothing parameter selection.

One might argue that theory should guide the amount of smoothing to be applied. And, in fact, theory is an excellent guide, if the theory is formulated at that level of detail. Many theories in the social sciences are articulated at a general level that may only predict a linear versus a nonlinear functional form. Therefore, most social science theories are not specific enough to provide any information about how smooth the relationship may be between x and y. Since most theories in the social sciences are silent about the smoothness of the process, theory is a poor guide for the selection of these model parameters. If theory does not provide any insight into the amount of smoothing to use, automated techniques are a reasonable alternative.

Analysts often use smoothers in contexts where we cannot reasonably expect them to have theories. For example, smoothed time terms are often used to model duration dependence in binary time series cross-sectional data (Beck, Katz, and Tucker 1998). In this context, it is unlikely that an analyst will have any theoretical priors on how smooth the estimate of time should be for these models. Moreover, a better fitting smoothing model will better approximate the nature of the temporal dependency. Automated smoothing provides the best means of modeling the amount of smoothing to impose on an estimate of temporal dependency. Smoothers are also important diagnostic tools and are used to smooth residual plots. For the smoothing of residual plots, it is better to use automated smoothing since the analyst is searching for trends, and there is little reason to guess about the optimal amount of smoothing to use.

Third, the confidence bands that are estimated after visual selection of smoothing parameters are misleading. For example, an analyst may be using a *lowess* smoother and plot four or five fits each with a different value for the span. Each fit may be slightly different but shows the same overall pattern. The analyst then selects one of the fits, and he or she estimates 95% confidence bands. Since they are typically estimated via pointwise standard errors and a normal approximation, these confidence bands do not in any way reflect the analyst's uncertainty as to which of the smoothed fits is best. Typically with automatic parameter selection, the confidence bands reflect the estimation uncertainty of the estimate for the smoothing parameter. This is not always strictly true. Some software for smoothing splines simply uses the standard pointwise confidence bands, but most newer software estimates the correct confidence bands. The estimate of λ provided by either the likelihood or cross-validation are subject to estimation uncertainty, and this estimation uncertainty is equivalent to the analysts' uncertainty over which smoothing parameter to choose. But the estimation uncertainty for λ will be incorporated into the estimated confidence intervals with automated smoothing, while the analyst's uncertainty over which smoothing parameter to use will not be reflected in the confidence bands with the visual method. Here, automatic parameter selection is superior to visual methods as it better captures our uncertainty about the level of smoothness.

Finally, the use of automatic parameter selection should not lead to blind adherence. Overfitting or rough fits can occur with automated smoothing. When such instances occur, a fit done with an automatic smoothing parameter algorithm should be compared to some fits with the smoothing chosen by visual inspection. Analysts should report if the automatic smoothing results differ dramatically from fits that use reasonable ranges of smoothness. Here, the automatic smoothing can serve as an important check on the analysts assumptions about how smooth the process is. We now provide several examples of using automated smoothing.

Automated Smoothing Examples

The first example uses simulated data which allows for an analysis of how well automated smoothing routines capture a true functional form. With real data, we never know the true functional form making it impossible to know how well a statistical model performs. For the simulation, we create a y that is a highly non-linear function of x. We compare how well two different smoothers estimate f, the relationship between x and y. For a more precise comparison, we calculate the mean squared error for each smoothed fit. In the analysis, we compare natural cubic B-splines and smoothing splines. For the natural cubic B-splines, we selected the number of knots with the AIC. For the smoothing splines, we use the mixed model spline smoother, so no input from the analyst is required.

The nonlinear functional form for y takes the following form:

$$y = \cos^2(4\pi x^3) \tag{4.8}$$

where

$$\varepsilon \sim N(0, 0.6). \tag{4.9}$$

The x variable is simply a sequence of 1000 values between 0 and 1. This function produces a highly nonlinear y, which may be difficult for the automated smoothing algorithm. We do not use any smoothers with the visual method in this example, since the high degree of nonlinearity in the function makes the visual method easy to use. The results of the first analysis are in Figure 4.4. The fit with natural cubic B-splines is in the first panel of Figure 4.4.

Knot selection via AIC does a poor job of estimating the smoothed fit between x and y. The fit is very poor for higher values of x as the spline fails to approximate the function at all. Stone *et al.* (1997) demonstrated the asymptotic equivalence between AIC and cross-validation, but these results suggest that AIC is an imperfect criterion for knot selection as large sample sizes are required for the two methods to produce similar results. The smoothing spline estimate is in right panel, and we see that even though the smoothing spline did not require any input from the analyst, it produces a good fit to the data. While some minor

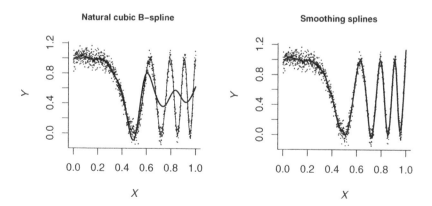

Figure 4.4   A comparison of estimated smoother fits.

Table 4.1   Mean squared error for smoother fits.

	MSE
Lowess	0.0052
Cubic B-splines	0.0469
Smoothing splines	0.0039

undersmoothing does occur, the overall fit is very faithful to the true functional form.

Table 4.1 contains the mean squared error for the two smoother fits, since the MSE allows for a more precise judgement of how well each method performs. For this exercise, we include the MSE for *lowess* fit with the visual method. Using the mean squared error criterion, the difference between *lowess* and smoothing spline fits is small, but the fit with smoothing splines using automated smoothing is slightly better. Moreover, one achieves this superior fit with no guesswork from the analyst.

In the next example, the data generating process for y is

$$y = \cos^4(4e^{x^3})   \tag{4.10}$$

where

$$\varepsilon \sim N(0, 1).   \tag{4.11}$$

The values for x and the sample size remain the same. This function produces a nonlinear relationship between x and y that is more difficult for the eye to discern. A scatterplot of the two variables is in Figure 4.5 along with the true functional form. With this set of simulated data, visual identification of the smoothness will be more difficult.

Figure 4.6 contains a plot of *lowess* fits with six different span settings and a second plot with a smoothing spline fit that uses the likelihood method. *Lowess* produces a variety of fits that all capture the true relationship between x and y. While one of the fits is obviously too rough and others are too smooth, two or three produce fairly reasonable fits that an analyst might reasonably select as the final fit. It is this uncertainty that will not be reflected in the estimated confidence bands. The smoothing spline fit very closely matches the true functional form, but does so with no input from the analyst, and the estimation uncertainty of λ will be reflected in the confidence bands. Finally, we plot the *lowess* fit with the lowest MSE against the smoothing spline fit in Figure 4.7.

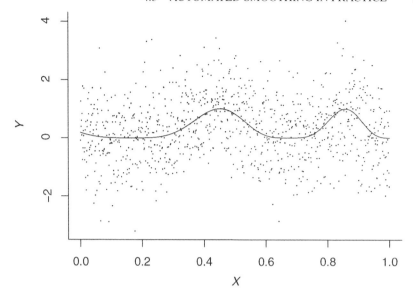

Figure 4.5   A simulated nonlinear smooth functional form.

Close examination of the *lowess* fit in Figure 4.7 with the lowest MSE (at 1.01) reveals that it is a fairly rough fit. This *lowess* fit is rough enough that we would want to increase the span according to the guidelines of the visual span selection. The smoothing spline fit with automated smoothing is smooth with a MSE that is only slightly higher (1.02) than the *lowess* model with the lowest MSE. Lowess fits that are smoother have MSE values that are higher than that of the smoothing spline. The analysis with simulated data demonstrates that smoothing spline fits with automatic smoothing parameter selection produce fits that closely match the true relationship between x and y. We now explore some empirical examples to demonstrate further automated smoothing.

Data Visualization

In statistics, there is a strong tradition of data visualization. Before statistical models are fit to the data, visualization is used to examine the structure of the data. Cleveland (1993) provides numerous examples where visual examination of the data reveals anomalies or features in the data that were missed in model-based analyses. Smoothers are integral to data visualization as a means of revealing trends in the data and helping the analyst better see the dependency between variables (Cleveland 1993). The goal of data visualization is to discover patterns, and automated smoothing techniques are a useful aid. An analyst may choose

Lowess – six different spans

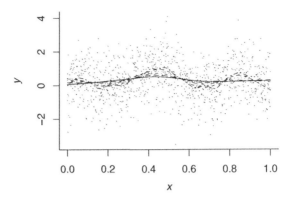

Spline – automatic smoothing

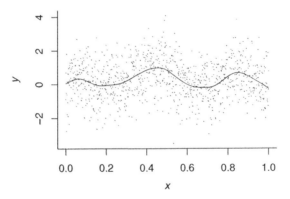

Figure 4.6   Comparing fits between smoothing splines and *lowess* with various spans.

to oversmooth a nonlinear feature in fear of overfitting. Checking a nonlinear pattern against an automated smooth can help the analyst to understand whether a feature may be real or the result of overfitting. Here, we provide two examples of data visualization using smoothers. The first example simply demonstrates that automated smoothing often produces results little different from visual methods. The second example indicates how automated smoothing can be used to check for overfitting.

Jacobson and Dimock (1994) study whether the number of overdrafts for a member of Congress in the House Banking scandal affected the challenger's vote share in the 1992 Congressional elections. We might expect a nonlinear

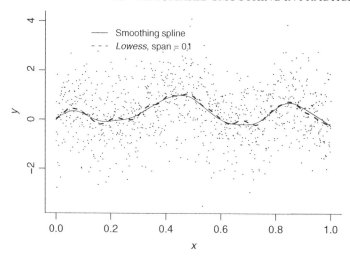

Figure 4.7   *Lowess* span chosen by MSE and smoothing spline fit.

dependency between these two variables as the challenger's vote share should increase as the number of overdrafts increase, but diminishing marginal returns may occur as additional overdrafts help the challenger less. Before estimating a parametric model, visual examination of a scatterplot between these two variables might reveal the suspected nonlinearity, and the addition of a smoothed fit will certainly help detect any nonlinearity. A smoother is well suited to estimating the relationship between the two variables without assuming the a specific functional form for the nonlinearity. A logarithm is often used to transform the data when diminishing marginal returns are suspected, but a logarithm imposes a form on the data, and since we can test the nonparametric regression fit against the logarithm, a spline is best for an exploratory analysis.

To demonstrate that using automatic smoothing techniques often produces a fit little different than fits an analyst might choose using visual methods, we estimated four different smoothing spline models between the challenger's vote share and the number of overdrafts. For three of the models, we manually selected the degrees of freedom, and for the fourth, we use GCV to estimate the smoothing parameter. Figure 4.8 contains the results from these four spline models. The upper-left panel of Figure 4.8 contains a plot where the amount of smoothing was estimated by GCV. First, for all the spline fits, we see an obvious nonlinear dependency between these two variables. While there are some minor differences, we observe the same pattern where the effect of the number of overdrafts on challenger's vote share levels off once a threshold is met. The similarity of

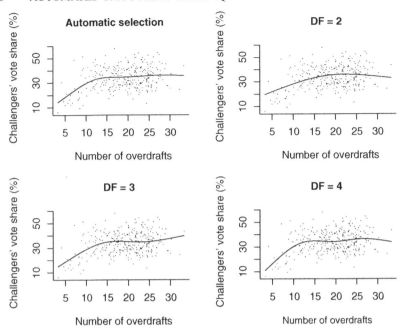

Figure 4.8   Manual selection of degrees of freedom compared to automatic selection for overdrafts and challenger's vote share.

the fits produces some amount of guesswork as one would be hard pressed to choose between the three manually selected fits. The automated smoothing model provides an answer that relies on the data.

If we were to add confidence bands for these fits, those for the smoothing spline with automated smoothing would reflect the estimation uncertainty for λ. For the other fits, the confidence bands would not capture our uncertainty about which degrees of freedom setting to use. This evidence suggests that automatic smoothing parameter selection produces fits that are equally smooth as user-selected fits but appeal to data instead of guesswork. Moreover, the smoothing spline fit would produce confidence bands that more accurately reflect our uncertainty.

In a second data visualization example, we use data from the 2000 US presidential election in Florida. These data are county-level counts of votes for Al Gore and the number of overvotes in each county. An overvote occurs when a voter votes for more than one candidate for a single office. The number of overvotes is not typically used to predict the number of votes a candidate receives, but given the irregularities in the 2000 Florida election, we might expect some dependency between these two measures. The data have some properties that suggest a transformation. The data are counts, bounded from above, and have an increasing variance, all of which suggests a logarithmic transformation. But the analysis of voting outcomes and election irregularities is not well developed in political science as the study of the topic originates with the election controversy in Florida in 2000. Given that the nature of election irregularities is not well understood, the immediate application of a parametric model is not advisable especially since we can test the spline model against a log transformation. Visualization allows for an examination of the data structure. Figure 4.9 contains three scatterplots of the data each with a different smoothing spline fit. One could also use a smoother with a log transformation to see if the relationship in nonlinear in the log scale as well.

As we discussed in Chapter 3, for smoothing splines without automated smoothing, four or five degrees of freedom is a typical starting point. The first spline fit in the top panel uses five degrees of freedom. First, we clearly observe a nonlinear dependency between votes and overvotes. It would appear that in counties where Al Gore received a higher number of votes the number of voting irregularities was higher. We might wonder whether the feature in the data that occurs between approximately 80 and 100 overvotes is real or perhaps overfitting. In the second panel of Figure 4.9, we decrease the degrees of freedom for the smoothing spline, and the feature disappears. Therefore, it is possible that the feature is the result of overfitting and not a real feature. One way to tell is to use an automated smoothing technique to see if the feature remains. The bottom panel in Figure 4.9 contains a smoothing spline done with GCV. We see that the feature remains, so the automated smoothing is useful tool for the data visualization in this illustration. In the first example, no differences emerged across the various fits, but with the Florida voting data, we find an unusual feature in the data. It is hard to know whether we should smooth this feature out or decide to consider it as something real. Automated smoothing provides another method to see if the data support such a feature.

Finally, we test the smoothing spline estimate against a parametric model with the data logged. The results from this test decisively indicate that the spline spline is preferred to the log fit ($p < 0.001$). The nonparametric model also provides a useful check on assuming that this data is best modeled with a log transformation.

Smoothing Time For Binary TSCS

One area of political science where splines are frequently used is in the analysis of binary time-series cross-sectional data. As Beck, Katz, and Tucker (1998) demonstrate, including a time counter on the right hand side of a logistic regression model will account for duration dependence in such data. While a number

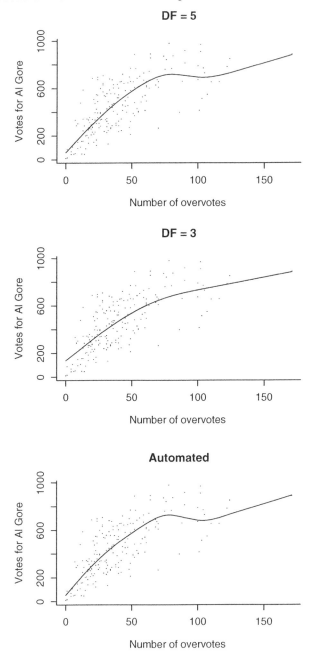

Figure 4.9   Scatterplot of county level votes for Al Gore and overvotes with smoothing spline fits, Florida 2000.

Table 4.2   Likelihood ratio test between cubic and smoothing splines.

T-statistic	16.48
p-value	0.003

χ^2, df ≈ 2

of functional forms for time are possible, they recommend using natural cubic splines for fitting such models. Of course, while natural cubic splines are a reasonable choice, they do require the analyst to choose the number of knots. As we have argued, knot selection is not difficult, but this is an area where analysts are are unlikely to have strong prior convictions about the number of knots required. The functional form of temporal dependence could be complex and perhaps highly nonlinear. Smoothing splines with automated smoothing, of course, do not require the analyst to make decision about knots.

We test whether using smoothing splines improves the overall fit of the model versus the knot selection used in Beck, Katz, and Tucker (1998). Performing such a test is straightforward. A model with cubic splines and a model with smoothing splines are nested which implies that a likelihood ratio test between the model with analyst-chosen knots and a model that uses smoothing splines will tell us whether there is any statistically significant difference between the two models.[3] For the test, we reestimate one of the logistic regression models with natural cubic splines with three knots from Beck, Katz, and Tucker (1998).[4] We estimated the same model except we use smoothing splines with GCV instead of natural cubic splines. We performed a likelihood ratio test to compare the fit across the two models. We do not present the specific results from the two models, which are immaterial, but the results from the likelihood ratio test are in Table 4.2.

Smoothing splines improve the overall fit of the model considerably, as we are able to reject the null that the extra parameters in the smoothing spline model are unnecessary at the 0.01 level. In this example, smoothing splines remove the uncertainty about the number of knots while also improving the overall fit of the model. For these types of models, automated smoothing is very useful. Temporal dependence may have a complex functional form, and the analyst rarely has a

---

[3]Technically, the test statistic distribution is nonstandard, but most texts argue that while the likelihood ratio test is only approximate, it is still reasonably accurate (Hastie and Tibshirani 1990; Ruppert, Wand, and Carroll 2003; Wood 2006). For information on more accurate tests, see Chapter 8 on bootstrapping tests.

[4]More specifically, we reestimated the logit model with splines from Table 1 from Beck, Katz, and Tucker (1998).

theory-based reason for choosing the level of smoothing. Automated smoothing allows for a data-based estimate of the smoothing.

Diagnostic Plots

One important step in the statistical modeling process is to check the plausibility of the model assumptions. One method for checking model assumptions is visual examination of residual plots. The eye can be deceived by what may or may not be a pattern in the plot, but smoothers can help reveal patterns and detect nonlinearities. With residual plots, the analyst is unlikely to have any theory to guide the degree of smoothing to impose. Moreover, oversmoothing can make a pattern appear linear when we are trying to detect nonlinearity. In such situations, it is better to rely on an automatic smoothing procedure and let the data, typically in the form of model residuals, speak for themselves.

There are a wide variety of diagnostic plots that analysts might examine, and for many, smoothers will be of little use. But for some model plots, a smoothed fit should always be added to the plot. One such example is model checking for the Cox proportional hazards model for duration data. Box-Steffensmeier and Zorn (2001) outline how a key assumption for the Cox model is that the hazard ratios are proportional to one another and that proportionality is maintained over time. One diagnostic that they recommend is the visual examination of the smoothed Schoenfeld residuals for the estimated model in question. For such plots, a smoother should always be added to aid in detection of nonproportional hazards.

For linear regression models, a smoother should always be added to plots of the model residuals against the fitted values.[5] A common variation on this plot is to use standardized residuals, where the raw residuals are divided by their estimated standard deviation. For such a plot, the residuals should be evenly scattered above and below zero. Systematic patterns in the variation of the residuals is a sign of heteroskedasticity. Using standardized residuals can make it easier to assess the constant variance assumption of least squares. For such plots, a smoother should also be added. A trend in the residuals is evidence of incorrect functional form or an omitted variable. When adding a smoother to such residual plots, automated smoothing can be helpful. Oversmoothing can result in flat fits which can be interpreted as a valid specification.

As an illustration, we use a data set from Li and Reuveny (2006). The authors argue that political factors such as regime type contribute to environmental degradation. They model a number of environmental indicators as a function

---

[5]Specifically, the residuals are $\hat{\varepsilon} = \mathbf{y} - \mathbf{X}\hat{\beta}$ and the fitted values are $\hat{\mathbf{y}} = \mathbf{X}\hat{\beta}$.

of democracy and autocracy scores, militarized conflict, gross domestic product (GDP) per capita, and population density. Figure 4.10 contains three plots of the standardized residuals against the fitted values for one of the authors' models.[6] In the top panel of Figure 4.10, we fit a smoothing spline model with three degrees of freedom. Here, one could reasonably assume that model is correct since there is a only slight departure from the 0 axis, but it appears to be minor. In the middle panel is a fit with four degrees of freedom, and the fit now has a slightly stronger trend. One might wonder: should additional degrees of freedom be used? Is there support for this trend in the data? The use of automated smoothing allows us to answer this question. In the bottom panel, we fit a smoothing spline model with the amount of smoothing selected by GCV. We now see an obvious trend in the plot, so it would appear that the model may be misspecified.

4.4   Automated Smoothing Caveats

Regardless of which smoother an analyst selects, he or she must always select the span, number of knots, or the degrees of freedom. Given the variance–bias trade-off involved in this choice, it can engender controversy, and undoubtedly, some of the controversy stems from the visual method of making this choice. The visual method can appear ad hoc, though, it should be noted that the visual method typically produces perfectly good results, but often there is no need to use the visual method. Smoothing splines with automatic smoothing parameter selection allow the analyst to estimate smoothed fits that rely on the data to select the amount of smoothing. Automated smoothing also produces confidence bands that better reflect our uncertainty about how smooth a process is. In many cases, automated smoothing techniques are useful and can be used with little concern. Unfortunately, automated smoothing techniques are not a panacea. If analysts have a theory about the level of smoothness or the results from automatic selection seem overly nonlinear or rough, they should select the amount of smoothing manually. Importantly, automated smoothing can be problematic in some contexts.

The estimate of λ through either GCV or a likelihood methods is subject to estimation uncertainty but improves as the sample size increases. Härdle, Hall, and Marron (1998) use simulations to demonstrate that as the sample size increases the estimates of the smoothing parameter from GCV converge to the true value quite slowly. The result is that, at times, the performance of automated techniques can be poor with smaller data sets (fewer than 50 or 100 cases). In small samples,

---

[6]Specifically, the plot is for the model of deforestation.

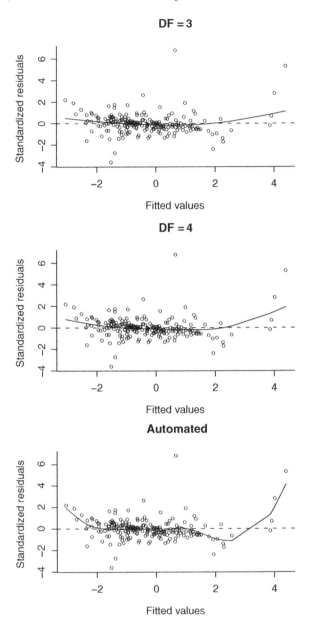

Figure 4.10   Standardized residual plots with smoothing spline fits.

the automatically smoothed fit may be overly jagged or highly nonlinear. In such situations, analysts will have to use the visual method.

Second, automated smoothing techniques have consequences for statistical inferences made with smoothers. Part of the power of smoothers stems from the ability to test whether a smoothed nonlinear fit is superior to a linear fit. As noted earlier, the referent F-distribution is only approximate for these tests. The use of automated smoothing techniques further complicates such tests. While work is ongoing in this area, it is known that the p-values for such tests will be lower than they should be. This is due to the fact that smoothing parameter uncertainty is not taken into account in the referent distributions. The general consensus is that when the p-value gives a clear-cut result it is safe to rely on the standard distributional tests, but if hypothesis tests are important, p-values near standard thresholds should be viewed with suspicion. In such situations, one should compare the linear model or the power transformation model to an overspecified *unpenalized* smoothing model such as natural cubic splines. Such a model will have the correct p-values (Wood 2006). The problem is not unique to smoothers, but is inherent to any model selection techinque. Importantly, analysts can use bootstrapping to conduct tests that are completely nonparametric allowing for comparisons between linear fits and nonparametric fits that use automated smoothing.

Barring these inferential or theoretical concerns, analysts are often better off letting the amount of smoothing be estimated from the data. Automatic smoothing removes any hint of art from the process, and more importantly provides confidence bands that more accurately reflect our uncertainty about the level of smoothness.

4.5   Exercises

Data sets for exercises may be found at the following website: http://www.wiley.com/go/keele_semiparametric.

1. For this exercise, use data on political demonstrations from Chapter 3.

   (a) Estimate a linear regression model where demonstrations is predicted by unionization and inflation. Plot the residuals against the fitted values and add a spline model. Experiment with user chosen values of smoothing and GCV selected smoothing? Does automated smoothing change your inference?

   (b) Use smoothing splines and the visual method to model the dependency between demonstrations and unionization. How many degrees of freedom do you select? Now add a spline fit using automated smoothing methods. How many degrees of freedom does the algorithm estimate are needed?

(c) Test a smoothing spline fit with GCV against a linear fit. Then test a model with natural cubic splines against a linear fit. Do the p-values differ across the two tests? Why might the p-values differ?

2. For this exercise, use the data on deforestation from Chapter 3.

(a) Model deforestation as a function of democracy and GDP using linear regression. Plot the residuals against the fitted values and add spline model with automated smoothing. Reestimate the model but now add a GDP squared term. Are the results from the residual plot improved?

(b) Test a smoothing spline fit with GCV against a linear fit for GDP and deforestation. Then test a natural cubic spline against a linear fit. Do the p-values differ across the two tests? Which p-value should you trust?

3. Repeat Exercise 1 using the data on wages from Chapter 3. Use age and wages for the second and third analyses.

5

Additive and Semiparametric Regression Models

Thus far the reader has been introduced to a variety of nonparametric regression models. The virtue of these models is that they allow the analyst to estimate the relationship between X and Y with few assumptions about the functional form. Nonparametric regression models, however, are unable to estimate multivariate specifications beyond those with two predictors, which seriously limits the utility of nonparametric regression in much applied research. The majority of all published quantitative research in the social sciences consists of statistical models with multivariate specifications with at least five predictor variables if not many more. The inability to estimate multivariate specifications would suggest nonparametric regression is unusable with the standard research designs in the social sciences. Under this scenario, nonparametric regression is a tool for exploratory analysis before proceeding to a multivariate specification.

With one additional assumption, however, we can add nonparametric regression to the standard multivariate models that are the stock and trade of quantitative research in the social sciences. If we are willing to assume that the effects of predictor variables are additive, we can form a semiparametric regression model where some of the covariates enter the model in a parametric fashion, while other variables enter as nonparametric terms. The semiparametric regression model allows the analyst to estimate standard parametric models with nonparametric estimates for continuous predictor variables and for easy diagnosis of nonlinearity within the context of the familiar multiple regression model. This chapter begins with a discussion of the additive model which is not a semiparametric but provides the basis for the semiparametric regression model. We then outline the semiparametric regression model along with methods of inference and estimation before proceeding to two empirical examples.

5.1   Additive Models

Recall that in Chapter 2, we discussed generalizing nonparametric models to multivariate specifications. In principle, we can specify the following general k-variate nonparametric regression model:

$$Y_i = f(X_1, \ldots, X_k) + \varepsilon. \tag{5.1}$$

In this form, one might assume that the nonparametric regression is suitable for the standard multivariate specifications used by social scientists, but as we discussed previously, this model has two serious limitations. First, interpretation is impossible when k exceeds 2. The output for Equation (5.1) when k equals 2 is a three-dimensional plot of a plane which represents the joint effect of the two predictor variables. While it is possible to estimate this model with a third predictor variable, the output would require a plot with an additional dimension, which is impossible.

Second, even if one could visualize the results, local estimation in k dimensions is problematic. Defining a local neighborhood for estimation is subject to the *curse of dimensionality* when k is greater than 1, since for a fixed number of points, local neighborhoods become less local as the number of dimensions increase. Bellman (1961), who coined this term, described a generic example of this dimensionality curse. For a unit interval [0, 1] with 100 points that are evenly spaced along this interval, the average distance between each point will be 0.01. For a cube where each side has the same unit interval length, 10^{20} points would be required to achieve a spacing of 0.01 between points on the surface of the cube. More generally, the curse of dimensionality demonstrates that the number of points required for even spacing of points along the surface of a cube is several orders of magnitude larger than for the same spacing along a line. How does this apply to local estimation?

Assume that the 100 points are data. If the 100 data points are distributed along the range of an X variable defined on the unit interval, there will be little space between the data points. In the context of nonparametric regression, this implies that the span can be quite small, and we will still have stable local estimates. But if those same 100 data points are distributed across a plane, the amount of space between the data points will be much larger, and the span will have to be much larger for a smooth estimate. Indeed, it is possible that to achieve a smooth fit that the span will have to be increased to the point that the fit will no longer be local but instead is global. As such, the curse of dimensionality defeats the very purpose of local estimation. The data in multiple dimensions becomes so sparse that the span must be increased to a point where the estimate is no longer local. In short, without very large amounts of data, it is impossible to get local estimates that are not highly variable.

Extending the nonparametric regression model to more than two X variables requires the additional assumption of additivity. While the assumption of additivity is more restrictive than a fully multivariate nonparametric regression model, it is among the most common assumptions for the parametric models used by analysts in the social sciences. For a parametric linear regression model, analysts typically assume the following functional form

$$Y_i = \alpha + \beta_1 X_1 + \cdots + \beta_k X_k + \varepsilon. \tag{5.2}$$

In this functional form, the analyst assumes that the effects of the X variables are additive on Y_i: the effect of X_1 and X_2 is $\beta_1 + \beta_2$. The additivity assumption is, at times, partially relaxed in the form of an interaction

$$Y_i = \alpha + \beta_1 X_1 + \beta_2 X_2 + \beta_3 X_1 X_2 + \varepsilon. \tag{5.3}$$

The additivity assumption is more restrictive but is typically already present in most models that we estimate and is readily extended to nonparametric regresssion models. A nonparametric regression model with additive effects has the following form

$$Y_i = \alpha + f_1(X_1) + \cdots + f_k(X_k) + \varepsilon \tag{5.4}$$

where $f_1, \ldots, f_k$ are arbitrary smooth but otherwise unspecified functions that we estimate from the data via a smoother. Nonparametric regression models with an additivity assumption are called additive models, and the motivation for the additive model stems from the desire to make nonparametric regression into a multivariate data analytic technique (Hastie and Tibshirani 1990).

The relatively routine assumption of additivity purchases the analyst much in terms of additional interpretability, since this assumption implies that multiple nonparametric regression models have the same interpretation as multiple regression models. For example, consider the following additive model

$$Y_i = \alpha + f_1(X_1) + f_1(X_2) + f_3(X_3) + \varepsilon. \tag{5.5}$$

Given the additivity assumption, we can plot each $\hat{f}_i$ individually since the effect of X_1 and X_2 is now $\hat{f}_1 + \hat{f}_2$. Moreover, the estimate of $\hat{f}_1$ takes into account the covariance between all three X variables and can be interpreted as the effect of X_1 on Y_i holding X_2 and X_3 constant. Now each $\hat{f}_i$ in the model is analogous to the β coefficients from a multiple regression model. Additivity allows us to retain the flexibility of a nonparametric regression model but provides parametric model like interpretation.

For all the advantages that additive models have over multiple nonparametric regression models, they still have limitations that make them generally unusable in the social sciences. Specifically, the additive model lacks one important refinement that is contained in the semiparametric regression model.

5.2   Semiparametric Regression Models

If we only estimated nonlinear relationships between continuous variables, the additive model would be perfectly suitable, but social scientific data analysis rarely conforms to such a paradigm. First, discrete predictor variables are common and often outnumber continuous predictors in a specification. Second, the relationship between Y and X may actually be linear. If a single parameter adequately captures the dependency between Y and X, there is no reason to expend the additional parameters required for a nonparametric regression estimate. In sum, the most flexible model will allow us to mix parametric terms with non-parametric terms in the same model. Fortunately, it is straightforward to alter the additive model to estimate both nonparametric and parametric terms. A model that mixes parametric and nonparametric terms takes the following form

$$Y_i = \alpha + f_1(X_1) + \cdots + f_j(X_j) + \beta_1 X_{j+1} + \cdots + \beta_k X_k + \varepsilon. \qquad (5.6)$$

In the above model, the first j covariates are assumed to have a nonlinear effect on Y_i and are fitted with nonparametric smoothers. The rest of the covariates enter into the model parametrically. Adding a parametric component to the additive model creates a *semiparametric* regression model, which provides the analyst with the ability to estimate a wide variety of functional forms. The parametric part of the model allows discrete covariates such as dummy variables or ordinal scales to be modeled alongside nonparametric terms, and any continuous covariates that the analyst deems to have a linear effect on Y_i can be estimated parametrically to save additional parameters. The semiparametric model also has an obvious advantage over a fully parametric model. In this semiparametric framework, the inferential machinery of nonparametric regression is retained in the semiparametric model, which allows the analyst to test whether any nonparametric terms are needed in the model specification. We can also include a variety of interaction terms in the semiparametric regression model. For example, consider the following semiparametric model:

$$Y_i = \alpha + f_1(X_1) + f_2(X_2) + \beta_3 X_3 + \beta_4 X_4 + \varepsilon. \tag{5.7}$$

In the above model, X_3 is a dummy variable and X_4 is continuous as are X_1 and X_2. With the semiparametric regression model, the analyst can specify several different interaction terms. First, the analyst can estimate a nonlinear interaction between X_1 and X_2. The result will be a three-dimensional plot identical to that from multiple nonparametric regression. The analyst can also include a completely parametric interaction term between X_3 and X_4. Furthermore, the analyst can mix interactions across the parametric and nonparametric terms in the model such that the nonlinear effect of X_1, for example, can be plotted at differing levels of either X_3 or X_4. In short, the semiparametric regression model is a versatile tool for modeling both linear and nonlinear dependencies between variables. We next outline the estimation process for both the additive and semiparametric regression models.

5.3   Estimation

Iterative algorithms are required for the estimation of both the additive and semiparametric regression models, and a variety of iterative algorithms have been implemented into various software platforms.[1] One method is to use the equivalence between smoothing splines and mixed models for a restricted maximum likelihood estimator that relies on a Newton–Raphson algorithm.[2] It is also possible to adapt the standard iterative reweighted least squares algorithm from the generalized linear model framework to estimate semiparametric models. The most well-known method of estimation for these models, however, is the backfitting algorithm proposed and implemented by Hastie and Tibshirani (1990). The backfitting algorithm is flexible enough to mix the estimation of both nonparametric and parametric terms and is also relatively simple.

5.3.1   Backfitting

Fitting models with multiple nonparametric terms is simple if the X variables in the model are uncorrelated. If the X variables are orthogonal, we can estimate each

---

[1] In fact, different algorithms are implemented in the same software package. For example, within R, the statistical software used for the analyses in this text, all three of the algorithms mentioned here have been implemented.

[2] For a detailed outline of this method see Ruppert, Wand, and Carroll (2003).

part with a series of bivariate models using OLS for the parametric components and *lowess* or splines for nonparametric components. In this context, there is no need to use multiple regression given that a set of bivariate regressions will be equivalent. In general, however, it is rare to have data where at least some of the predictor variables are not correlated, and hence we need a method to estimate the terms from an additive or semiparametric model that accounts for the covariances between predictor variables. The backfitting algorithm is designed to take these correlations into account when estimating nonparametric and parametric terms.

The backfitting algorithm is suggested by the idea of a partial regression function; consider the following two predictor additive model:

$$Y = \alpha + f_1(X_1) + f_2(X_2) + \varepsilon. \tag{5.8}$$

Assume we know the true form of f_2 but not f_1. If this were true, we can rearrange the equation to form the partial regression function in order to solve for f_1:

$$y - \alpha - f_2(X_2) = f_1(X_1) + \varepsilon. \tag{5.9}$$

Smoothing $y - \alpha - f_2(X_2)$ against X_1 produces an estimate of $f_1(X_1)$. Therefore knowing one of the partial regression functions allows us to estimate the other partial regression function. In truth, we don't actually know either of the regression functions, but if we assume a set of starting values for one of the f terms, the partial regression functions suggest an iterative solution for the estimates in an additive model. We wish to estimate the following additive model:

$$Y_i = \alpha + f_1(X_1) + \cdots + f_k(X_k) + \varepsilon. \tag{5.10}$$

In Equation (5.10), let $\mathbf{S}_j$ denote a matrix where each column represents each estimate of f_k, and $\mathbf{X}$ be a model matrix where each column is one of the X covariates. The backfitting algorithm for the estimation of additive models has the following steps:[3]

1. First, set $\alpha = \bar{Y}$ and $\mathbf{S}_j = \mathbf{X}$ as starting values for $j = 1, \ldots, m$.

2. The algorithm loops over the columns of $\mathbf{S}_j$ to calculate a partial residual for each X variable. Below is the estimate of the partial residual, $\hat{e}_p^j$ for X_1:

$$\hat{e}_p^j = Y_i - \sum_{i=2}^{k} \mathbf{S}_j - \alpha \tag{5.11}$$

where the second term on the right side represents row sums across $\mathbf{S}_j$ for the X variables $k \geq 2$.

---

[3]This algorithm is one described by Hastie and Tibshirani (1990).

3. Smooth e_p^j on X_1. The analyst must choose a nonparametric regression model for this step along with the span or degrees of freedom for the smoother depending on the choice of nonparametric regression model. (Most software implementations use splines for the reasons outlined in Chapters 3 and 4.)

4. Replace the covariate X_1 in $\mathbf{S}_j$ with the predictions from the smoothed fit at the values of X_i.

5. Repeat steps 2–4 for each X from 2 to k.

6. Calculate the model residual sum of squares as:

$$\text{RSS} = \sum_{i=1}^{n} \left[\left(Y_i - \sum_{i=1}^{k} \mathbf{S}_j \right)^2 \right] \tag{5.12}$$

where we again sum across the rows of $\mathbf{s}_j$.

7. If the change in the RSS is within a specified tolerance level, the model has converged and the algorithm stops. If not, the process repeats until the change in the RSS is within the specified tolerance level.

Once the algorithm stops each column of $\mathbf{S}_j$ contains the nonparametric estimate of each X variable on Y. Importantly, these estimates now take into account the covariances between the X variables, so if one were to estimate an additive model with three X variables, the plot of $\hat{f}_1$ is interpreted as the effect of X_1 on Y_i, holding X_2 and X_3 constant. A number of variations on the basic backfitting algorithm described above are possible; such as using OLS estimates as starting values. This approach is outlined below:

1. Estimate the following mean deviated linear regression:

$$Y_i - \bar{Y} = \beta_1(X_1 - \bar{X}_1) + \cdots + \beta_k(X_k - \bar{X}_k) + \varepsilon$$
$$Y^* = \beta_1 X_1^* + \cdots + \beta_k X_k^* + \varepsilon. \tag{5.13}$$

The parameters $\beta_1 \cdots \beta_k$ serve as starting values for the iterative backfitting algorithm.

2. The partial residual for X_1 is estimated:

$$\hat{e}_{px_1} = Y^* - \beta_2 X_{i2}^* + \cdots + \beta_k X_k^*. \tag{5.14}$$

The estimation of the partial residual removes the linear dependence between Y and X_2 but the linear relationship between Y and X_1 is retained along with any nonlinear relationship in the least square residuals ε, for $j = 1, \cdots, m$.

3. Next, the partial residual is smoothed against X_1 providing an estimate of f_1:

$$\hat{f}_{x_1} = \text{smooth}[e^j_{px_1} \text{ on } X_1]. \tag{5.15}$$

(Again, the smoother used in this step is largely immaterial.)

4. The estimate of $\hat{f}_{X_2}$ is formed from the partial residual for X_2:

$$e_{px_2} = Y_i^* - \hat{f}_{x_1} X_1^* + \cdots + \beta_k X_k^*. \tag{5.16}$$

5. We now smooth this partial residual against X_2 which provides an estimate of $\hat{f}_{X_2}$:

$$\hat{f}_{x_2} = \text{smooth}[e_{px_2} \text{ on } X_2]. \tag{5.17}$$

6. The new estimate of $\hat{f}_{X_2}$ is used to calculate an updated partial residual for X_3. Once there is an initial estimate for each f_k term, the process repeats.

7. This iterative process continues until the estimated partial-regression function stabilizes and the fit of the function as calculated by the change in the residual sum of squares is with a specified tolerance level.

When the process is done, we will have estimates for the partial effects of X on Y_i. That is, the algorithm returns the smoothed estimate of X_1 on Y_i controlling for the effect of $X_2, \ldots, X_k$.

The backfitting algorithm for the semiparametric regression model proceeds in a similar fashion. The partial residual is formed for each predictor variable, and if the the analyst has selected a particular covariate to have a nonlinear fit, the partial residual for that variable is smoothed on that variable. If the variable is selected to have a parametric estimate, the partial residual is regressed on that variable using least squares instead of a smoother. This modification allows the backfitting algorithm to estimate semiparametric regression models. If the backfitting algorithm is applied to a linear model and linear regression replaces the nonparametric regression in the estimation step, it will produce least squares estimates.[4] Despite its advantages, backfitting does have one flaw; it is difficult to incorporate automated smoothing techniques into the algorithm. Newer software implementations of semiparametric regression models often use iterated reweighted least squares algorithms which do not have this deficiency. See Wood (2006) for a description of these algorithms.

---

[4]Hastie and Tibshirani (1990) provide descriptions of several other modifications that can be made to the basic backfitting algorithm.

5.4   Inference

Statistical inference for semiparametric regression models mixes inference for linear models with inference for nonparametric regression models. For the nonlinear terms in the models, we would like to estimate confidence bands and test the model fit against more parsimonious parametric specifications. For the parametric terms in the model, we would like estimated standard errors to form confidence intervals and to perform hypothesis tests. All of these are available with the semiparametric regression model.

Confidence bands for the nonparametric terms and standard errors for the parametric terms require an estimated variance–covariance matrix; estimation of the variance–covariance matrix is similar to that for nonparametric regression models but considerably more complex. Testing the semiparametric regression model against fully parametric alternatives, however, is straightforward using either an F-test or a likelihood ratio test. We start with the derivation of the variance–covariance matrix for the estimates from the backfitting algorithm, which we use to construct confidence bands for the nonparametric terms and standard errors for the parametric terms. Next, we describe the procedure for hypothesis tests using either partial F-tests or the likelihood ratio test.

Confidence Bands and Standard Errors

Recall that in Chapters 2 and 3, we defined $\mathbf{S}$ as the smoother matrix, which is analogous to the hat matrix $\mathbf{H}$ from linear regression, such that $\hat{\mathbf{f}} = \mathbf{S}\mathbf{y}$. Once we obtained $\mathbf{S}$, standard errors were estimated in a fashion similar to least squares, using $\hat{\sigma}^2 \mathbf{S}\mathbf{S}'$ as the variance–covariance matrix. We need to construct a variance–covariance matrix for additive and semiparametric regression models which will allow for the same inferential methods. Using L_2 function spaces, Hastie and Tibshirani (1990) demonstrate that the additive model can be written as the following set of equations, where $\mathbf{I}$ is an $n \times n$ identity matrix and $\mathbf{S}_1, \ldots, \mathbf{S}_k$ is the smoother matrix for each X variable. The constant can be omitted with no loss of generality:

$$
\begin{bmatrix}
\mathbf{I} & \mathbf{S}_1 & \mathbf{S}_1 & \ldots & \mathbf{S}_1 \\
\mathbf{S}_2 & \mathbf{I} & \mathbf{S}_2 & \ldots & \mathbf{S}_2 \\
\vdots & \vdots & \vdots & \ddots & \vdots \\
\mathbf{S}_k & \mathbf{S}_k & \mathbf{S}_k & \ldots & \mathbf{I}
\end{bmatrix}
\begin{bmatrix}
\mathbf{f}_1 \\
\mathbf{f}_2 \\
\vdots \\
\mathbf{f}_k
\end{bmatrix}
=
\begin{bmatrix}
\mathbf{S}_1\mathbf{y} \\
\mathbf{S}_2\mathbf{y} \\
\vdots \\
\mathbf{S}_k\mathbf{y}
\end{bmatrix}.
\tag{5.18}
$$

In theory, the above set of equations can be solved directly using a noniterative procedure such as QR decomposition, but this system of equations is typically too large for such noniterative methods, necessitating the use of an algorithm. We can rewrite the above equation more compactly and rearrange it to form a matrix analogous to a hat matrix:

$$\hat{\mathbf{S}}\mathbf{f} = \hat{\mathbf{Q}}\mathbf{y}$$

$$\hat{\mathbf{f}} = \hat{\mathbf{S}}^{-1}\hat{\mathbf{Q}}\mathbf{y}$$

$$\hat{\mathbf{f}} = \mathbf{R}\mathbf{y}, \qquad (5.19)$$

where $\mathbf{R} = \hat{\mathbf{S}}^{-1}\hat{\mathbf{Q}}$. The matrix $\mathbf{R}$ is a linear mapping of $\mathbf{y}$ to $\hat{\mathbf{f}}$, and for additive and semiparametric models, it is the analogue to the hat matrix. If the observations are independent and identically distributed then:

$$\mathbf{V}(\hat{\mathbf{f}}) = \sigma^2 \mathbf{R}\mathbf{R}' \qquad (5.20)$$

where σ^2 is replaced by

$$\hat{\sigma}^2 = \frac{\sum e_i^2}{df_{\text{res}}}. \qquad (5.21)$$

The residual degrees of freedom are $df_{\text{res}} = n - \text{tr}(2\mathbf{R} - \mathbf{R}\mathbf{R}')$. Confidence bands can be constructed using ± 2 times the square root of the diagonal elements of $\hat{\sigma}^2 \mathbf{R}\mathbf{R}'$. For semiparametric models, the diagonal elements of $\mathbf{R}$ are also the variances for any β's estimated. Estimation of $\mathbf{R}$ requires an iterative method. For example, Hastie and Tibshirani (1990) apply the backfitting algorithm to compute $\mathbf{R}$, but this estimate of the variance–covariance matrix does not correct for bias in $\hat{\mathbf{f}}$. We can again correct for bias using either the mixed model formulation of splines or Bayesian confidence bands. Either method introduces further complications which require careful attention to numerical stability in the programming of the estimation algorithms. Both Ruppert, Wand, and Carroll (2003) and Wood (2006) provide details on the estimation of bias adjusted variance–covariance matrices for additive and semiparametric regression models.

Hypothesis Tests

Hypothesis testing for additive and semiparametric regression models does not present any complications. For parametric terms, the standard errors from $\mathbf{R}$ allow for the usual inferential procedures such as hypothesis tests. For the nonparametric terms there are two hypothesis tests of interest both of which are conducted with either a partial F-test or likelihood ratio test. The first test indicates whether the effect of X on Y is significantly different from zero, whereas the second test establishes whether the nonparametric fit for a variable improves the model fit over the use of a parametric term. Demonstration of both of these tests is best done through a generic example. Consider the following model:

$$Y = \alpha + f_1(X_1) + f_2(X_2) + \varepsilon. \tag{5.22}$$

To test whether the effect of X_2 is statistically different from zero, we would test the model above against

$$Y = \alpha + f_1(X_1) + \varepsilon. \tag{5.23}$$

To test whether f_2 is linear, we would test Equation (5.22) against

$$Y = \alpha + f_1(X_1) + \beta_1 X_2 + \varepsilon. \tag{5.24}$$

The F-test is based on the residual sum of squares. Define the residual sum of squares for any additive or semiparametric regression model as

$$\text{RSS} = \sum_{i=1}^{n}(y_i - \hat{y})^2. \tag{5.25}$$

Let RSS_0 be the residual sum of squares for the restricted model while RSS_1 is the residual sum of squares from the additive or semiparametric model. The test statistic is

$$F = \frac{\text{RSS}_0 - \text{RSS}_1/[\text{tr}(\mathbf{R}) - 1]}{\text{RSS}_1/df_{\text{res}}}. \tag{5.26}$$

This test statistic follows an approximate F-distribution with $df_{\text{res,smaller}} - df_{\text{res,larger}}$ and $df_{\text{res,larger}}$ degrees of freedom (Hastie and Tibshirani 1990). For software that uses re-weighted least squares or restricted maximum likelihood instead of backfitting for estimation, we use either a likelihood ratio or difference of deviance test. The likelihood ratio test for additive and semiparametric models takes the usual form:

$$\text{LR} = -2(\text{LogLikelihood}_0 - \text{Loglikelihood}_1) \tag{5.27}$$

where Loglikelihood_0 is the log-likelihood for the restricted model and Loglikelihood_1 is the log-likelihood for the unrestricted model, the additive or semiparametric regression model. The test statistic under the H_0 follows an approximate χ^2 distribution, and the degrees of freedom is the difference in the number of parameters across the two models. The deviance for a model is simply -2 times the loglikelihood. The test statistic for a difference of deviance test is the difference of the two model deviances. The resulting test statistic still follows a χ^2 distribution, and the degrees of freedom for the test remain the difference in the number of parameters across the two models.

One caveat with respect to inference is in order. If automated smoothing techniques are used in the estimation of the nonparametric terms, the distributions for the test statistics are thought to approximately follow a χ^2 distribution, but it is less well understood how close the approximation is (Wood 2006). In general, we can expect the estimated p-value for the hypothesis tests to be too small. If hypothesis tests are of interest and the p-value allows one to narrowly reject the null hypothesis, manual smoothing selection should be used. With splines, if the p-value is of particular interest, unpenalized splines with a surfeit of knots will provide the most accurate p-values. The analyst can also use bootstrapping to conduct these tests without reference to a parametric distribution. See Chapter 8 for details.

5.5   Examples

This section contains two detailed examples of how one might apply semiparametric models to social science data. In many ways, these models are similar to nonparametric regression models, but the main difference is that the analyst estimates one or more nonparametric models within the framework of a linear regression model. Estimation of a semiparametric model requires a few basic steps. First, the analyst must choose which variables will be nonparametric terms in the model. For the nonparametric fits, one can either use automatic smoothing or choose the amount of smoothing to be applied to each predictor variable that is to be estimated nonparametrically. Finally, the analyst estimates the model. Now one can inspect plots, examine the parametric estimates, and test hypotheses.

One final note: local polynomial regression smoothers are not typically used for the nonparametric component of semiparametric models. Most newer semiparametric software does not even include an option for these smoothers, hence, in the examples that follow, we use smoothing splines to estimate the nonparametric terms.

5.5.1   Congressional Elections

We return to the 1992 House elections data from Jacobson and Dimock (1994). Thus far, I have only examined the dependency between one predictor, support for Perot, and the vote share of challengers. Of course, Jacobson and Dimock estimate a more complicated model with a multivariate specification. More generally, Jacobson and Dimock (1994) are interested in testing whether challengers in the 1992 House elections received a boost from the fallout of the House Banking scandal. To test this, they estimate a linear regression model where the challenger's vote share is a function of whether the they had political experience, their campaign spending, incumbent spending, presidential vote in that district for the last presidential election, the number of bad checks that the incumbent wrote on the House Bank, the margin of victory for the incumbent in the last election, whether the district had been redistricted, and the percentage of votes that Ross Perot received in that district in the 1992 presidential election.

Additive Models

We start by estimating a simple additive model with two predictors. In Chapters 2 and 3, we saw that the nonparametric estimate between the support for Perot and the challenger's vote share was nonlinear and that the nonparametric estimate was superior to a linear regression fit with either a logarithmic or quadratic transformation. While this model revealed the nonlinear dependency between the two variables, it did not account for other possible predictors that may affect the challenger's vote share. An additive model allows us to estimate the effect of support for Perot and with other continuous predictors. In this illustration, we add the overdrafts measure to the model and estimate a model with the following specification

$$\text{Challenger's vote share} = \alpha + f_1(\text{Perot}) + f_2(\text{Overdrafts}) + \varepsilon. \quad (5.28)$$

This additive model is estimated with smoothing splines and GCV for automated smoothing and the results are presented in Figure 5.1. The left panel of Figure 5.1 is the plot of $\hat{f}_1$, the nonparametric estimate of the effect of support for Perot on challenger's vote, holding the overdrafts variable constant. Interpretation of the plot is the same as for any spline smoother; we have a visual representation of how the challenger's vote share changes as support for Perot increases. The effect of overdrafts also appears to be nonlinear as well in that the challenger's vote share climbed as the number of overdrafts increased, but once the number

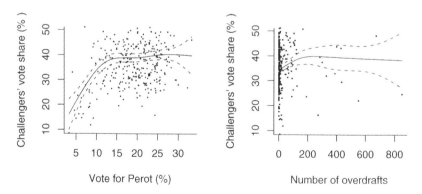

Figure 5.1   Additive model for challenger's vote share as a function of support for Perot and the number of overdrafts.

of overdrafts was greater than 200, it did little further damage to the incumbent. Therefore, the marginal cost of 200 overdrafts differed little from 600 overdrafts.

In contrast to our analysis, Jacobson and Dimock logged both of these variables in their models. The inferential machinery of nonparametric regression remains intact for additive models, which allows us to test whether the the additive model fit improves upon linear model fits with and without the log transformation used by the authors. To illustrate this, we first test the additive model against a linear regression model where the variables for Perot support and overdrafts are untransformed. The data clearly allow us to reject the null hypothesis of a linear model using a likelihood ratio test. The χ^2 test statistic of 2428.2 on 4.49 degrees freedom is well below the critical values for conventional levels of statistical significance (p-value < 0.001). Next, we test the additive model against a model with logs. While we reject the log transformation for the Perot variable, the test indicates that spline fit does not improve upon the logarithmic transformation for the overdrafts variable ($p = 0.30$). Given the test results, one can opt for a more parsimonious specification for the overdrafts variable, though one might still prefer the nonparametric estimate for this variable. The reason for this is that the logarithmic transformation requires that we arbitrarily add some positive number to each observation to avoid taking the logarithm of zero which is undefined, while the spline model requires no such arbitrary transformation of the data. While the additive model is a clear improvement over nonparametric regression, we would prefer the flexibility of the semiparametric regression model, which will allow us to include categorical predictors.

A Semiparametric Model

With the additive model, however, we cannot include discrete predictors or model continuous predictors parametrically, but the semiparametric regression model allows us to include both types of predictors. The measures of whether the district was competitive or not in the last election, whether the district underwent a partisan redistricting, and whether the challenger is experienced are all dummy variables and therefore need to be estimated as parametric terms in the model. For the remaining continuous covariates, we need to decide (1) whether the effects are nonlinear and (2) whether data transformations provide adequate approximations for any nonlinearity. In the original model estimated by Jacobson and Dimock (1994), three variables were logged: challenger spending, incumbent spending, and overdrafts. We test all the continuous predictors both for nonlinearity and whether common transformations are adequate.

The specification of the semiparametric model is necessarily an iterative process as the analyst must decide which of the continuous covariates should be modeled parametrically and which should be modeled nonparametrically. Standard practice is to first estimate a model with all continuous covariates estimated nonparametrically. An examination of the plots from this model will often clearly reveal which terms can be modeled with a linear fit. Next, this fully nonparametric model can be tested against a series of more restrictive models, and the best procedure is to test the fully nonparametric model against a model where one of the continuous predictors is either linear or transformed, repeating the process for each continuous variable. After this series of tests, the analyst can choose a final specification. We followed this procedure using the Jacobson and Dimock data, testing spline fits against linear functional forms and models with logarithmic and quadratic transformations.

The results from the model comparison tests are in Table 5.1. First, the linear terms are inadequate on all five accounts given that for each variable, a nonparametric fit better is than a linear fit. This is not the case for some of the models with a logarithmic transformation. For the overdrafts and presidential vote variables, the spline estimates are no better than a logarithmic transformation, so one might opt for a parametric term. When we compare the spline models to quadratic fits, we find that only the spline estimate for challenger spending is superior to models with quadratic transformations. Given these results one might model all of the variables, with the exception of challenger spending, with either log or quadratic transformations. Despite such test results, one might still use the spline fits here. The only reason to use models with transformations is on grounds of parsimony. The smoothing spline fits, however, often use nearly the same degrees of freedom as a quadratic fit. For example, the difference in the degrees of freedom for the model with incumbent spending estimated with a quadratic term and the smoothing spline fit is 0.08. In this example, the spline fits add approximately one degree of freedom over a model with quadratic terms. If one has a reasonably large data set, the difference of a single degree of freedom may be of little consequence.

Table 5.1 Testing spline fits against linear fits and transformations: Challenger's vote share.

	Linear	Logarithmic	Quadratic
Challenger spending	<0.001	0.004	<0.001
Incumbent spending	0.013	0.02	0.20
Challenger's party's	<0.001	0.10	0.10
Presidential vote overdrafts	0.046	0.30	0.10
Perot support	<0.001	0.01	0.10

Note: p-value for test of semiparametric model against parametric model.

The semiparametric model provides a framework for testing assumptions about the functional forms of multivariate models. As a first step, the analyst can simply plot the spline fit for any continuous variables, and this visual test typically makes it readily clear whether the functional form is nonlinear. After the visual test, the analyst can then test the spline fits against a linear model for more precision. Finally, the spline fit can also be tested against a variety of data transformations to see if they allow for a more parsimonious specification. Furthermore, transformations are not always possible due to the scale of the variable, and in such situations, the analyst can still use a spline. Finally, the results from testing may reveal that a fully parametric model is an acceptable functional form. If so, splines are unnecessary, but now the analyst can justify the assumption of a parametric model.

With the functional form tests completed, we now consider the differences between the semiparametric model and the parametric model. Recall that incorrect functional forms are a form of misspecification that can lead to biased estimates. It is often the case that once we model nonlinearity, the model results change. For example, a variable may appear to be statistically significant, but once we model the nonlinearity, we see the effect was spurious due to misspecification. Of course, the reverse can be true as well. We illustrate this point by replicating the fully parametric model used by Jacobson and Dimock (1994).[5] First, we estimated two different semiparametric models where in the first, we modeled the support for Perot nonparametrically, and in the second, we modeled support for Perot parameterically. For both semiparametric models, the overdrafts variable is modeled parametrically but is logged. For the other continuous variables, we opt for spline fits over quadratic transformations given that the gains in parsimony are small. The results for the three different models are in Table 5.2.

The model in the first column replicates the results in Jacobson and Dimock (1994). In this model, we find that five of the variables have effects that are statistically significant at the traditional 0.05 level. First, consistent with conventional analyses of Congressional election outcomes, presidential vote share and challenger spending were strong predictors of the challenger's vote. Support for Perot and the number of overdrafts, two factors that were specific to the 1992 election, also helped challengers. We might be skeptical about the results of this model, however, given that the functional form is incorrect, and the likelihood ratio test indicates that the semiparametric model in the second column provides a significantly better fit to the data ($p < 0.001$).

---

[5]The results, here, are quite close to those originally reported by the authors, but differ slightly due to some data errors that were fixed by the authors after publication. The basic inferences remain the same across the models.

Table 5.2   A comparison of parametric and semiparametric models of the 1992 House elections

	Challenger vote share (%)	Challenger vote share (%)	Challenger vote share (%)
Experienced	0.86	0.72	0.52
Challenger	(0.86)	(0.85)	(0.82)
Challenger	2.69***	_***	_***
Spending	(0.26)		
Incumbent	−0.02	–	–
Spending	(0.60)		
Presidential	0.36***	_***	_***
Vote	(0.04)		
Log overdrafts	0.83*	0.86***	0.85***
	(0.19)	(0.18)	(0.18)
Competitive in 1990	0.87	0.67	0.68
	(0.83)	(0.79)	(0.79)
Redistricting	3.26*	2.61*	2.59*
	(1.21)	(1.16)	(1.16)
Perot vote	0.18**	–	0.10
	(0.06)		(0.60)
Constant	3.93	33.02*	31.15*
	(3.47)	(0.81)	(1.39)
R^2	0.60	0.65	0.65
LR test p-value		0.00	0.10

Likelihood ratio test against previous model in the table.

Standard errors in parentheses. Two-tailed tests.

* p-value < 0.05 ** p-value < 0.01 *** p-value < 0.001

This result is not surprising for several reasons. In the original analysis, the variables for challenger and incumbent spending were logged, but as the likelihood ratio tests in Table 5.1 demonstrated, splines provide a better fit for these variables. The log transformation assumes a particular functional form for the nonlinearity, and the spline model allows us to test this assumption. For the spending variables, the log assumption is incorrect, but for the overdrafts variable it proved to be the correct transformation. There also several substantive differences across the models. Most notably in the parametric model, the effect of presidential vote is assumed to be linear as a one percentage point increase in presidential vote share corresponds to a 0.36 percentage point increase the challenger's vote share, plus or minus 0.08 points. How does this compare to the

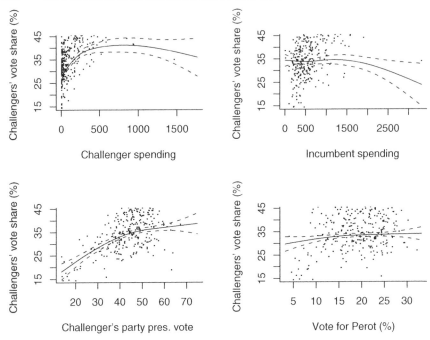

Figure 5.2   Nonparametric estimates for semiparametric regression model of Congressional challenger's vote share, 1992.

nonparametric estimate? As demonstrated in Figure 5.2, which contains plots of the four spline fits from the semiparametric model along with biased corrected 95% confidence bands, there is a threshold effect for presidential support and support above 50% does little to aid the challenger. The nonparametric estimate is also revealing about the nature of candidate spending as the slope for challenger spending is very steep before it reaches a threshold of around $500 000. Clearly, spending by the challenger below this threshold has a strong effect on vote share, and a logarithmic transformation obscures this insight.

The reader should also notice that estimates for incumbent spending and support for Perot are essentially flat. Statistical significance tests confirm that the 95% confidence intervals for these variables are not bounded away from zero, which leads to another difference between the semiparametric and parametric models. In the parametric model, support for Perot appeared to be a predictor of support for the challenger, but once we correctly account for nonlinearity in the model, the effect proves to be spurious. This is an important point; using the correct functional form can often change the conclusions we draw from a model.

Here, once we account for the nonlinearity in the model, the effect of support for Perot is no longer statistically significant, and in the plot, the effect looks linear. In the third column of Table 5.2, I reestimate the model with the effect of support for Perot as linear. Again, we find no effect, and this model fits the data no better than one with a spline estimate for support for Perot.

This example highlights the flexibility and power of semiparametric regression models, which provide a framework for testing whether a linear functional form is justified. Moreover, the analyst can also test whether standard transformations such as quadratic terms are adequate models of nonlinearity. Finally, we find that modeling the nonlinearity changes our inferences. In the parametric model, support for Perot appears to be a key predictor of support for the challenger, but once we take nonlinearity into account, we find that Perot support seems to be unrelated to support for the challenger.

Interactions

Thus far, we relaxed the linearity assumption but retained the additivity assumption for the Congressional elections model, but it is not unusual to relax the latter assumption with interaction terms. As indicated earlier in this chapter, we can include interactions in semiparametric regression models. In the semiparametric regression model, we can either interact a nonparametric term with a discrete variable or interact two continuous variables which allows for an estimate of the jointly nonlinear effect of the two variables. Examples of both follow.

Suppose we suspect that the effect of support for Perot depends on whether the congressional district was competitive in 1990. Jacobson and Dimock included a predictor of district competitiveness, which they measured with a dummy variable indicating whether the incumbent won by more than 10 percentage points in the last election. Using this measure, we want to test whether the marginal effect of support for Perot is conditional on whether the district was competitive or not, and testing this hypothesis requires an interaction between these two variables. In this instance, we model the interaction between one variable that is fit nonparametrically and a discrete variable since the competitiveness measure is a dummy variable. For this interaction, the model output will be a nonparametric plot for each level of the discrete modifier variable, and therefore, the semiparametric estimate for the interaction will be two plots: one plot of support for Perot in competitive districts and another plot of support for Perot in noncompetitive districts. We added this interaction to the semiparametric model estimated in the last section, and the results are in Figure 5.3. In the left panel, we see there is little relationship between support for Perot and the challenger in districts that were competitive in 1990, but in the second panel, we see a strong nonlinear

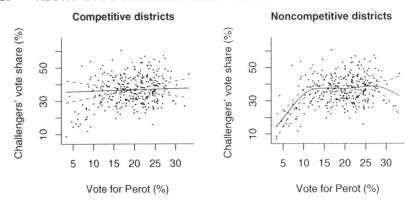

Figure 5.3   Interaction between support for Perot and competitive districts.

dependency in noncompetitive districts. Clearly, the effect of Perot support is conditional on the type of district.

The nonparametric component of the semiparametic model also allows the analyst to model nonlinear interactions between continuous variables. Here, the visual nature of interpretation for nonparametric terms is useful. For example, Jacobson and Dimock (1994) included an interaction between the challenger's spending and overdrafts in one of their models, since they suspect that the scandal was not enough to bring down an incumbent; the implication being that unless a well financed challenger spent money to educate voters about the incumbent's financial profligacy, the scandal might not have mattered to the incumbent's reelection chances. In the semiparametric regresion model, we can include an nonlinear interaction between these two variables, and the output from such a specification will be a plane in the form of a three-dimensional plot that allows the analyst to examine the conditional effects of each variable. We reestimated the full model for the challenger's vote share but specified a nonlinear interaction between challenger spending and the number of overdrafts. We omit the full model results, but the plot of the interaction is in Figure 5.4.

The plot reveals several interesting patterns. First, we see that the effect of overdrafts is lower when challenger spending is low, as the plane slopes upward across the range of challenger spending. The plot also reveals that a moderate number of overdrafts still had potent effects even when the challenger spent little money. Second, the effect of overdrafts is much closer to being linear for challengers that spent the most money. Besides capturing the nonlinearity, such plots are a useful means for reporting interactions as they convey the conditional effects implied by an interaction.

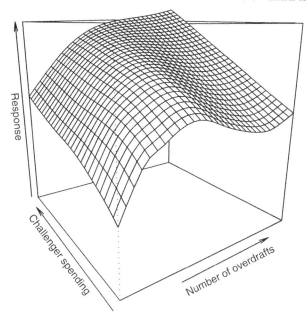

Figure 5.4   Interaction between challenger spending and the number of overdrafts.

Finally, we discuss two last aspects of on nonparametric interaction terms. First, one can also plot confidence planes for these plots, but without color, it is difficult to distinguish between the estimate and the confidence planes. Second, nonlinear interaction terms present no complications for testing between models. The likelihood ratio allows us to test between the model with a nonlinear interaction and a semiparametric regression model with a standard interaction between the two variables. The test indicates that the fit of a model with the nonlinear interaction is superior to a semiparametric regression model without the interaction ($p < 0.01$).

5.5.2   Feminist Attitudes

Quantitative analyses of survey data are common in the social sciences. In the next example, we analyze survey data from a study of the determinants of feminist attitudes using data from the General Social Survey. Bolzendahl and Myers (2004) compare interest-based and exposure-based explanations of feminist opinions. Under interest-based explanations, it is assumed that when a person's defined interests benefit from gender equity they will be more likely to support feminist attitudes. For exposure-based explanations, it is assumed that people develop feminist attitudes when exposed to ideas and life experiences that reinforce such attitudes. In their analysis, they regress various scales of feminist opinions on labor force participation, the proportion of household income a woman earns, hours worked, spousal work status, measures of family structure, socialization, and several demographic measures.

Many of the regressors in this model are discrete and are not suited to non-parametric modeling, but there are three continuous regressors, which implies that we need to test for nonlinear dependencies between these variables and the response. In the analysis conducted by Bolzendahl and Myers (2004), the three continuous variables are: hours worked, age, and proportion of income earned by the respondent. The authors estimated models for four different feminist attitudes in two different time periods with separate models for men and women. In the analysis that follows, we replicate the model of attitudes toward family responsibilities for the second time period (1987–2000) using only women. The full model specification has 23 predictor variables, so we only focus on the three continuous covariates. While we fit the full model as specified by the authors, we do not present the results for the parametric portion of the model. For this specific model, Bolzendahl and Myers (2004) find that age and hours worked have statistically significant effects on feminist attitudes while the proportion of income earned by the woman did not matter.[6]

We begin by estimating a semiparametric regression model using smoothing splines fits for the three continuous predictors using GCV for smoothing parameter selection. Before conducting formal tests, we plot the three nonparametric estimates since visual inspection of the fits can readily reveal any nonlinearity. Figure 5.5 contains plots of the nonparametric fits for these three variables, and a cursory inspection is informative for all three estimates. First, the fit for the proportion of income earned appears to be highly nonlinear, and there are also noticeable undulations that might be the result of overfitting. Hours worked appears to have little impact on feminist attitudes, as the estimate is nearly flat. Finally, age appears to be a strong influence, but it appears to be linear. This illustration demonstrates the usefulness of visual testing, which suggests the use of manual smoothing for one variable and testing the spline fit for age against a parametric fit. We estimated a second model removing the smoothed terms for age and hours worked. A likelihood ratio test indicates that the fit for the age

---

[6]The proportion of income measure is simply the woman's reported income divided by reported family income. This measure does not have a maximum value of one as we might expect. The authors do not comment on this anomaly, which is presumably caused either by overreporting of individual income or underreporting of family income.

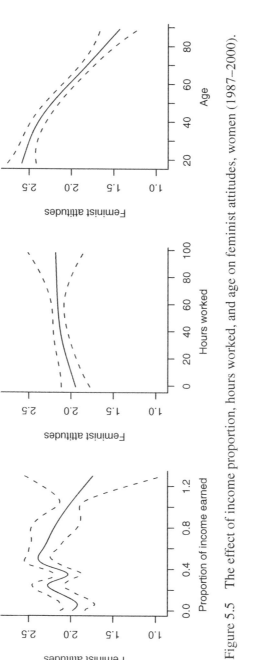

Figure 5.5　The effect of income proportion, hours worked, and age on feminist attitudes, women (1987–2000).

variable is better when modeled with a smoothing spline as the χ^2 test statistic is 58 on 2.01 degrees of freedom, so despite what we observe the spline fit is superior to a parametric estimate.

We take a moment to further consider the possible overfitting in the estimate for income proportion. In the plot, several undulations in the nonparametric fit for income proportion are clearly visible. It might be tempting to ascribe some meaning to this highly local nonlinearity, but one should avoid the temptation. The data set is large (nearly 6000 cases), which can make idiosyncratic variation in the nonparametric estimate more common; in such situations, it is best to manually select the amount of smoothing. We reestimated the model using four degrees of freedom for the smoothing parameter on that variable only, and the results are in Figure 5.6. As the reader can see, reducing the amount of smoothing eliminates the wavy features in the estimate. A likelihood ratio test indicates that the model with manual smoothing selection is still a better fit to the data than a parametric estimate. This example demonstrates that while automated smoothing techniques are often reliable overfitting is still possible, and analysts must use manual smoothing to correct such overfitting when it occurs.

Finally, we consider whether any misspecification occurred due to unmodeled nonlinearity, and we find that the inferences from the semiparametric model differ from the fully parametric model. In the semiparametric model, the effect for income proportion is now highly significant ($p < 0.001$), while it failed to reach statistical significance in the parametric model. Moreover, the effect of

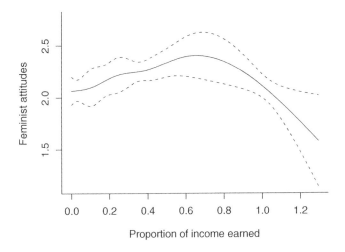

Figure 5.6   The effect of income proportion on feminist attitudes with manual smoothing selection.

hours worked now fails to achieve statistical significance whereas before the estimate was bounded away from zero. This is obvious in the plot as the effect of hours worked is essentially flat, while there is a strong downward slope for income proportion. A likelihood ratio test indicates that the semiparametric model provides a superior fit: the χ^2 test statistic is 257 on 8.14 degrees of freedom.

The semiparametric regression model in this example serves to illustrate two points. First, again we see how misspecification from the wrong functional form resulted in a number of different inferences. Income proportion became an important predictor once nonlinearity was taken into account. Moreover, once we modeled the nonlinearity, the effect of hours worked failed to achieve statistical significance, whereas it was significant in the fully parametric model. The example is also useful since it demonstrates that even with smoothing splines and automated smoothing overfitting can occur. Overfitting is more likely with large data sets such as this one, but reducing the degrees of freedom removes the overfitting. Importantly, though, the overfitting did not change the basic form of the estimate.

5.6   Discussion

Adding nonparametric components to parametric regression models produces a semiparametric regression model that can estimate both highly nonlinear functional forms and parametric estimates for discrete predictors. The semiparametric regression model retains the testing mechanism of nonparametric regression, which allows analysts to test whether parametric fits are adequate or not. The semiparametric regression model is a good compromise between the flexibility of nonparametric regression and the interpretability of parametric models.

While there are many benefits to using semiparametric regression models, certain caveats are in order. All of the pitfalls of nonparametric regression apply to additive and semiparametric models. First, when the smoothing parameter is estimated with an automated method, the test statistics for the likelihood ratio test remain approximate. If the test is borderline, the model can be reestimated using manual selection or using standard splines. The analyst can also simulate the p-values for the test. See Chapter 8 for details. Second, semiparametric models, just like parametric regression models, are also subject to multicollinearity. In semiparametric models, this is called concurvity. If two X variables are highly correlated, the backfitting algorithm may be unable to find a unique curve, and the result will be a linear fit. There is no solution short of additional data.

5.7   Exercises

Data sets for exercises may be found at the following website: http://www.wiley.com/go/keele_semiparametric.

1. The data in `allmen1986.dta` contains the same data used in the example on feminist attitudes. That analysis was restricted to women only and this data set contains the same set of variables for men.

   (a) Estimate a linear regression model with the data using the family responsibilities variable as the dependent variable. Produce partial residual plots for the continuous variables in this model. Add a smoother to these plots. Can you diagnose any nonlinearity from these plots?

   (b) Replicate the analysis in the second example using this new data set. Estimate the model with penalized splines and automated smoothing. Using likelihood ratio tests determine which model appears to fit the data best. Do the results here match those from the partial residual plots?

   (c) Do the results differ from the analysis here? Do the income proportion and age variables still have nonlinear effects? Does the automated smoothing algorithm appear to overfit the estimate for income proportion? If so, experiment with different smoothing parameters until the fit appears reasonable.

   (d) Re-estimate the model using unpenalized splines. Repeat the likelihood ratio tests. Do the p-values from the tests differ when compared to those for the models fit with smoothing splines and automated smoothing? Why might we expect them to differ? Which set of p-values are the most accurate?

   (e) Repeat the analyses above using the abortion attitudes variable as the outcome variable.

2. The `forest.dta` file contains data on evironmental degradation across countries. For the first model use the following variables as predictors: `dem`, `wardum`, `rgdpl`, `openc`, and `popdense`.

   (a) Fit a semiparametric regression model to the data using spline fits for all the continuous predictors. Plot all the nonparametric estimates. Which appear to be nonlinear?

   (b) Now use a series of likelihood ratio tests to decide which variables have nonlinear effects.

(c) For those variables that appear to be nonlinear, test whether power transformations model the nonlinear adequately.

(d) Decide on a final specification for the model. Estimate this model and provide an interpretation for what you find.

(e) Fit an interaction between democracy and GDP. Does there appear to be a nonlinear interaction between these two variables? Use a likelihood ratio test to determine whether the nonlinear interaction should be included in the model.

3. The data set `latamerturnout.dta` contains data on voter turnout from all minimally democratic elections from 1970-95 in Latin America. Develop a model of turnout. Decide which terms should be modeled parametrically and which should be modeled nonparametrically. Be sure to examine the appropriate model diagnostic plots. Compare your final semiparametric model to a parametric model. Do the results change? How does the use of the nonparametric terms change the substantive interpretation for those variables compared to parametric estimates?

4. Use the data set `forest.dta` to develop your own backfitting algorithm. Fit an additive model with two predictor variables. Use one of the algorithms outlined in this chapter. Compare your results to those from another software package. Do the estimates differ?

6

Generalized Additive Models

So far we have discussed semiparametric estimation solely in the context of linear regression models, but we can estimate a wide range of regression models semiparametrically. This chapter starts with a reprise of the generalized linear model (GLM) framework and then adapts it to semiparametric techniques to derive generalized additive models (GAMs), the semiparametric version of GLMs. We outline the estimation of GAMs as well as methods for statistical inference. We conclude with a number of examples. This chapter contains many examples due to the wide range of GLMs that we can use in conjunction with semiparametric techniques.

6.1 Generalized Linear Models

In the social sciences, categorical data are widespread, and models for categorical data analysis such as logistic regression and count models are common. In statistics, the models for such categorical data are referred to as generalized linear models (GLMs) (McCullagh and Nelder 1989). The GLM paradigm uses the basic framework of the linear regression model and adapts it to categorical dependent variables. Below we provide a brief outline of the GLM framework since it helps illuminate the need for semiparametric techniques for regression models with discrete outcome variables.[1]

In the GLM framework, models have three components: the stochastic component, the systematic component, and a link function. The analyst must choose each part depending on the substantive problem at hand. First, we select the stochastic component of the model, which is the sampling distribution for the dependent variable. Standard distributions for the stochastic component are the binomial, multinomial, Poisson, and negative binomial, though many others are possible. Second, we specify the systematic component. The most common form for the systematic component is

$$\eta = \mathbf{X}\boldsymbol{\beta}. \tag{6.1}$$

Of course, this is simply a linear and additive functional form. Third, we specify, $g(\cdot)$, the link function which links the stochastic and systematic components of the model by mapping Y_i onto the linear predictor through a mathematical transformation. The link function must be monotonic and differentiable to ensure a mapping between Y_i and the linear predictor η. Common link functions are the logit, probit, log, and identity. The GLM framework provides a common notation for both the linear model and nonlinear models such as logistic and Poisson regression models.

As we noted in Chapter 1, the GLM notation is useful since it clarifies the source of the nonlinearity in our models. This point is important enough, however, that it bears repeating. The basic functional form for GLMs is still linear and additive. The model predictions are linear in the systematic component until the link function is applied. For example, consider a probit model. The predictions from a probit model are linear until the Normal cumulative distribution function is applied to convert it to a nonlinear probability scale.

Of course, analysts can alter the systematic component to relax the assumptions of linearity and additivity. Introducing an interaction term adds a multiplicative component to the functional form. Importantly, if the true model is nonlinear in the variables, this nonlinearity must be modeled in the systematic component. The link function does nothing to model such nonlinearity. For example, if the true form of the systematic component is

$$g(\eta_i) = \alpha + \beta_1 X_{i1} + \beta_2 X_{2i}^2, \tag{6.2}$$

a quadratic term must be included on the right-hand side of the model regardless of what link function the analyst selects. As such, specification errors in the

---

[1] The material that follows including the notation is heavily dependent on McCullagh and Nelder (1989). Any reader interested in knowing more about GLMs should consult this classic text.

form of unmodeled nonlinearity are as likely with models that many analysts consider to be nonlinear. In fact, for most models within the GLM framework specification errors are more serious. With linear regression models, specification errors only affects variables that are correlated. But for many GLMs with nonlinear link functions, the variables need not be correlated for specification error to affect the estimates (Yatchew and Griliches 1985). Thus, regardless of what link function an analyst selects, he or she must still account for any possible nonlinearity in the systematic component of the model. Again, nonparametric techniques are useful for modeling nonlinearity in the systematic component of a GLM.

This brings us to generalized additive models (GAMs), which extend the generalized linear model in the same way additive models extend the linear regression model. To specify a GAM, we allow the linear predictor from a GLM to be some smooth function estimated from the data. For example, with a standard GLM we would write:

$$g(\eta_i) = \alpha + \beta_1 X_{i1} + \beta_2 X_{2i} \tag{6.3}$$

where $g(\ )$ is a link function chosen by the analyst. We can rewrite this GLM as the following GAM:

$$g(\eta_i) = \alpha + f_1(X_{i1}) + f_2(X_{2i}). \tag{6.4}$$

Here, we have replaced the parameters in the linear predictor with two smooth functions to be estimated from the data. The additivity assumption is retained to allow for interpretability of the nonparametric estimates. We can relax the additivity assumption in a limited way to model nonlinear interactions. With GAMs, we retain the testing framework which allows us to test the nonparametric estimates against linear fits or fits with power transformations. Like the additive model, the GAM is of limited usefulness unless it is altered to include parametric terms, but this presents no difficulties. In short, the semiparametric model from the last chapter can be extended to models for discrete data. A semiparametric approach to modeling discrete data will again provide a flexible framework for diagnosing and modeling nonlinearity. Before proceeding, we discuss a matter of notation. While GAMs are technically not semiparametric models, the term GAM is often applied to semiparametric GLMs. In the discussion that follows, we adopt this convention. Therefore, we use the term GAM to refer to both additive GLMs and semiparametric GLMs.

6.2   Estimation of GAMS

The estimation of GLMs is done either through direct maximization of the likelihood function using a Newton–Raphson algorithm or through iterated reweighted least squares (IRLS) (Hardin and Hilbe 2007). The Newton–Raphson method and its variants are probably more familiar to social scientists, but in statistics IRLS is commonly used to estimate GLMs. For interested readers, Hardin and Hilbe(2007) provide a useful outline of each method. IRLS is well suited to the estimation of GAMs since it reduces the estimation of GLMs to the iterative application of weighted least squares. This implies that we can use the backfitting algorithm outlined in Chapter 5 to estimate the smoothing components in GLMs. GAMs can also be estimated using Newton–Raphson methods, though it is rare, since it tends to be less stable than using backfitting. GAMs that rely on mixed model based splines use Newton's method. See Ruppert, Wand, and Carroll (2003) for an outline of how this estimation procedure is applied to GAMs.

Below is a basic sketch of the IRLS algorithm typically used to estimate GAMs. See Hastie and Tibshirani (1990) and Wood (2006) for a more detailed description of applying the IRLS algorithm to GAMs. In the example below, the model is additive but the algorithm can be adapted for semiparametric estimation.

1. Initialize estimated values α, a constant, and $f_1, \ldots, f_k$. Zeros are frequently used as starting values as are least squares estimates.

2. Calculate η_i, the linear predictor, by evaluating $\alpha + \sum_{j=1}^{k} f_k(X_k)$. Note that the summation is across the rows of the $\mathbf{X}$ matrix.

3. Calculate fitted values, μ_i, by evaluating $g(\eta_i)$. For example, with a Poisson regression model this would be $\mu_i = e^{\eta_i}$.

4. Construct an adjusted continuous dependent variable based on the linear predictor, fitted values and the outcome variable:

$$z_i = \eta_i + (Y_i - \mu_i) \left(\frac{\partial \eta_i}{\partial \mu_i} \right). \tag{6.5}$$

5. Assemble the following matrix of weights:

$$\Sigma_{ii} = \left(\frac{\partial \eta_i}{\partial \mu_i} \right)^2 (V_i)^{-1} \tag{6.6}$$

where V_i is the variance of y at μ_i.

6. Fit a weighted additive regression model using z_i as the response and Σ as the weights to get estimates for $f_1, \ldots f_k$. The backfitting algorithm is used in this step to estimate the weighted additive model.

7. The estimates $f_1, \ldots, f_k$ become a new set of starting values in step one, and the process repeats until the change in the estimates is small relative to a convergence criterion.

   The advantage of the IRLS approach is that estimating a GAM is not much more difficult than the estimation of an additive model. The estimation of GAMs is clearly more computationally intensive than an additive model. Consider that for the additive model, we only had to let the backfitting algorithm run once. Now we must estimate an additive model for each iteration of the IRLS algorithm. While modern computers may slow down a little when estimating a GAM, with large data sets the delay in estimating the model can be noticeable.

6.3   Statistical Inference

The basics of inference with GAMs are unchanged from additive models and semiparametric models. The variance–covariance matrix produces standard error estimates for confidence bands for nonparametric terms and hypothesis tests for parametric terms. Other hypothesis tests are done through model comparisons of likelihoods.

Confidence Bands

Another advantage of IRLS is that estimation of the variance–covariance matrix remains based on $\mathbf{R}$, the additive fit operator outlined in Chapter 5. For estimation purposes, the GAM is an additive model that uses a constructed outcome variable and weights, and this simplifies estimation of the variance–covariance matrix. We, first, define the covariance of the estimated linear prediction, $\hat{\eta}$. Since the GAM is equivalent to a weighted additive model, the following is true

$$\hat{\eta} = \mathbf{R}\mathbf{z} \tag{6.7}$$

where $\mathbf{R}$ is defined as in Chapter 5 and $\mathbf{z}$ is the vector of continuous outcomes constructed from the linear predictions, fitted values, and the response. This implies that the variance–covariance matrix for a GAM is

$$\mathrm{Var}(\hat{\eta}) \approx \mathbf{R}\Sigma^{-1}\mathbf{R}' \tag{6.8}$$

where Σ is the estimated set of weights from the IRLS algorithm. In the above equation, $\approx$ means 'asymptotically equivalent', which is true if the usual regularity conditions hold. This approximation is required since $\mathbf{R}$ is dependent on Y_i and μ in the IRLS algorithm, and therefore, it is no longer a linear operator (Hastie and Tibshirani 1990). With the estimated variance–covariance matrix, we form 95% confidence bands in the usual way: $\hat{f}_k \pm 2\sqrt{\text{Var}(\hat{\eta})}$.

Hypothesis Tests

Hypothesis testing for GAMs changes little from the last chapter. The only difference is that now we exclusively use the likelihood ratio test. The test is defined as:

$$LR = -2(\text{LogLikelihood}_0 - \text{Loglikelihood}_1) \tag{6.9}$$

where LogLikelihood_0 is the estimated log-likelihood from the more restrictive parametric model and Loglikelihood_1 is the estimated log-likelihood from the more general semiparametric model. Under the regularity conditions, the test statistic from the above likelihood ratio test is χ^2 distributed with degrees of freedom equal to the difference in the number of parameters across the two models. As before, we use the likelihood ratio to test for nonlinearities in the systematic component of the model by testing smoothed fits against a linear or transformed fits.

The degrees of freedom for GAMs are calculated from the trace of the $\mathbf{R}$ matrix as it was for the additive model. The residual degrees of freedom are

$$df_{\text{err}} = n - \text{tr}(2\mathbf{R} - \mathbf{R}'\Sigma\mathbf{R}\Sigma^{-1}). \tag{6.10}$$

For two nested models, the difference between their respective df_{err} values provides the degrees of freedom for the likelihood ratio χ^2 test.

One caveat should be offered. The distributional theory for the likelihood ratio test above is underdeveloped. It is not known whether the distribution for the test statistic is actually χ^2 distributed. Hastie and Tibshirani (1990) provide simulation evidence that the distribution of the test statistic is relatively close to a χ^2 distribution. The use of automated smoothing techniques further clouds the distributional theory. Despite this uncertainty, Hastie and Tibshirani (1990) argue that the χ^2 is useful as an informal method for comparing models. There are other methods for comparing GAMs. The simplest is to fit a GAM and plot the results. If the fit looks close to linear or quadratic one can then use a more parsimonious reparameterization. A more formal solution is to use bootstrapping to perform a nonparametric model comparison. Details on bootstrap model comparisons are provided in Chapter 8.

6.4   Examples

In the rest of the chapter, we present a number of examples. GAMs encompasses a variety of models, so we attempt to provide coverage of the more common GLMs that can be estimated as semiparametric models. We start with a logistic regression example.

6.4.1   Logistic Regression: The Liberal Peace

For the first illustration, we use data on international disputes. In international relations, much research focuses on the factors that promote or inhibit militarized conflict. While a large literature has developed on this topic, much of the focus is on whether democracies are less likely to engage in conflict with other democracies and whether trading partners are less likely to engage in conflict with each other. Many researchers have provided evidence for these two propositions (Russett 1990, 1993; Maoz and Russett 1992, 1993; Oneal *et al.* 1996; Oneal and Russett 1997). In general, it has been found that even controlling for a variety of confounding factors, the foundations for the 'Liberal Peace' findings remain intact, suggesting that democracy and trade are the best guarantors of peace.

We use a data set developed by Oneal and Russett to investigate the possibility of nonlinear effects. The data set is composed of 827 'politically relevant' interstate dyads for the period from 1950 to 1985. Each observation is a dyad year. A dyad is a pair of states, so for a single year each observation is a pair of states for all states in that year. For this time period, there are 20 900 dyad year observations with an average of 25.4 years per dyad. The outcome variable is the time until the onset of a militarized interstate dispute between the two nations that make up the dyad (Oneal and Russett 1997; Reed 2000). In past work, seven different factors have been identified as important to the risk of a dispute.[2] The seven factors are: (1) the level of *democracy* in the dyad as measured by a scale of Polity III scores, (2) *economic growth*, (3) the presence of an *alliance* between the two nations in the dyad, (4) geographical *contiguity* in the dyad, (5) the ratio of military *capability* between the two nations, (6) the level of intradyadic *trade* measured as a proportion of GDP lagged one year, (7) a counter for the number of *previous disputes* within the dyad, and (8) a counter for the number of *peace years* for the dyad.[3] I omit dyad-years of continuing conflicts as a means of accounting for repeated events. Of the variables above, four

---

[2] The operationalization of all the variables is identical to that of Beck, Katz, and Tucker (1998) as I use the same data set.

[3] Beck, Katz, and Tucker (1998) show that smoothing this counter helps account for duration dependence in the data.

are continuous, and thus might have nonlinear functional forms. For example, we might suspect that once some level of interdyadic trade is reached, further trade will no longer reduce possibility of conflict. The same may also be true for democracy.

We estimated a logistic GAM with smoothing applied to the measures of democracy, trade, economic growth, the capability ratio, and peace years. At this stage, we could either conduct a series of likelihood ratio tests or plot the four nonparametric estimates and inspect them for nonlinearity. Visual inspection of the plots may be enough to understand which terms can be modeled parametrically. Figure 6.1 contains plots of the four nonparametric estimates. The visual test is quite clear; three of the predictors have obviously linear functional forms. The reader may have noticed that the scale for the y-axis in Figure 6.1 is unusual. There are two reasons for this. First, all the variables are mean deviated by the estimation algorithm to increase numerical stability. This is why all the effects are centered at 0 on the y-axis. Second, we plotted the nonparametric estimates in the untransformed scale of the linear predictor, and therefore,

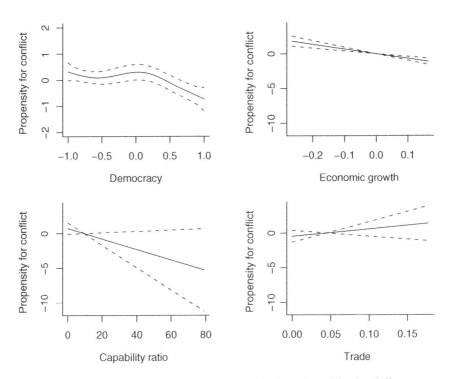

Figure 6.1   The nonparametric estimates for risk of a militarized dispute.

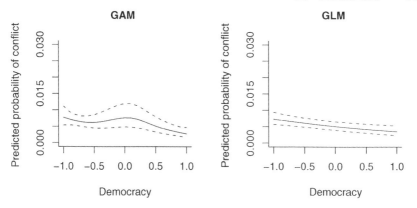

Figure 6.2   Comparison of the effect of democracy on militarized conflict: probability scale.

the predictions for Y are not on a probability scale. It is standard practice to convert logistic regression parameter estimates to odds ratios or predicted probabilities, and the same can be done for the nonparametric estimates, which we will do shortly. However, for visual diagnosis of nonlinearity, it is best to plot in the linear scale, since applying the link function can obscure the form of the nonlinearity.

Next, we used a likelihood ratio test to decide whether the effect of democracy was significantly nonlinear. The test indicates that the spline estimate for the effect of democracy is superior to using a parametric term. The final specification uses a spline fit for democracy but the other terms are modeled parametrically. We tested this specification against a model which was fully parametric. The model with a spline fit for democracy improves upon the fully parametric model ($p < 0.03$).[4] In this final specification, democracy remains statistically significant. Many of the control variables also have substantive effects. Economic growth and the presence of an alliance reduce the likelihood of conflict, while contiguity and past conflicts both increase the risk of conflict. As Beck, Katz, and Tucker (1998) found, trade does not seem to inhibit conflict.

In Figure 6.2, we plot the effect of democracy again, but now it is plotted on a probability scale. The nonlinear effect of democracy is now more readily apparent. Along the autocratic range of this scale, the probability of a conflict is roughly constant, though it does increase some around the middle of the scale.

---

[4]The fully parametric model is not actually fully parametric in that we smoothed the peace years variable to account for duration dependence.

Once past zero, however, there is a steady decline in the probability of a conflict as the level of democratization increases. The reader might also notice the scale for the y-axis. Holding the rest of the model constant, the probability of a militarized conflict is less than 0.01 for even the most autocratic of nations. Figure 6.2 also contains the estimate for democracy from a parametric model. We see that here the probability of conflict declines across the entire range of the democracy measure, so the using a fully parametric specification obscures the propensity of some autocratic nations to engage in conflict.

Converting the nonparametric estimate to the scale of the link function is done in an identical fashion converting to β's to this scale. Holding other covariates in the model constant at some appropriate category or value, we calculate the predicted value of Y for each value of the nonparametric estimate. We convert these predicted values that are in the scale of the linear predictor to a probability scale using the link function.

6.4.2   Ordered Logit: Domestic Violence

Ordered categorical variables are common in the social sciences, and either ordered probit or ordered logit models can be used to model such variables. We use data on violence between couples from Demaris *et al.* (2003) and Demaris (2004) to demonstrate the estimation of GAMs with discrete ordered outcomes. In both analyses, the analysts investigate the factors that are related to violence in domestic partnerships. They use a three category response variable that records the level of violence between couples over the time span of the relationship. The first category is reported intense violence, the second category is reported physical aggression, and the final category is an absence of reported violence. In the analysis, the following independent variables are used to predict the level of violence: a dummy variable for whether the couple are cohabiting as opposed to married, a dummy variable for a minority couple, female's age at union, the degree to which the male was isolated from his partner's relatives, the degree of economic disadvantage, a dummy variable for the presence of alcohol or drug abuse, the relationship duration in years, which has been mean-centered, a scale that measures the level of open disagreement, and a scale of communication style. The data are from Waves 1 and 2 of the National Survey of Families and Households and contains responses from 4095 couples (Sweet, Bumpass and Call 1988).

In his analysis, Demaris (2004) considers whether the effect of relationship duration is perhaps nonlinear, and he uses a quadratic transformation to account for the nonlinearity. In the analysis, the quadratic term is statistically significant so the effect is assumed to have a quadratic functional form. While statistical significance for the quadratic term is a clear indication of a nonlinear effect, the analyst has assumed the functional form of the nonlinearity is quadratic. While the quadratic transformation might be appropriate, without using the Box–Cox technique, it cannot be easily tested against other nonlinear forms besides higher order polynomials. For example, perhaps a logarithmic transformation would be a better alternative? It is also possible that no standard power transformation captures the nonlinear functional form.

Besides the measure of relationship duration, there are six other continuous measures in the model. We start the analysis by testing each one for nonlinearity. We find that among the seven continuous predictors, only the measure of relationship duration is nonlinear. We then estimate a GAM with a spline for only the relationship duration variable, which we test against a fully parametric ordered logit GLM. The likelihood ratio indicates that the GAM is a substantially better fit to the data than the GLM ($p < 0.001$).

Figure 6.3 contains a plot of the nonparametric estimate for relationship duration. The estimate appears to be a standard threshold effect where once relationship duration reaches approximately 10 years, the likelihood of a decrease in violence levels off. Recall that the relationship duration variable was mean-centered in the original analysis which accounts for the scale of the x-axis in the figure. In Figure 6.3, we plotted the effect in the scale of the linear predictor,

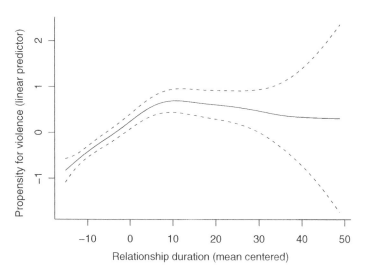

Figure 6.3   The effect of relationship duration on couple violence: linear predictor scale.

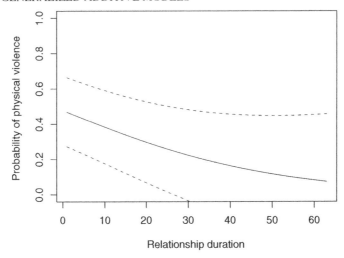

Figure 6.4   The effect of relationship duration on couple violence: probability scale.

which is useful for a visual inspection of the nonlinearity but should not be used for substantive interpretation unless the latent variable is rescaled to a unit variance and the coefficients are standardized (Long 1997). Instead, the analyst must convert the nonparametric estimate to a probability scale. Figure 6.4 contains a plot of the effect of relationship duration on the probability that there was physical violence in the partnership. Once other terms in the model where held constant and the estimate was transformed to the probability scale, the effect is still nonlinear and consistent with the basic finding but does not resemble the nonparametric estimate plotted in the scale of the linear predictor.

Thus far in the analysis, we have not tested whether a power transformation of the relationship duration variable might be appropriate. In the original analysis, a quadratic transformation was used to model the nonlinearity, but we might also consider a logarithmic transformation. The relationship duration variable is bound from below at 0 and may have a long tail for the upper bound, and logarithmic transformations are useful for producing symmetric distributions when data are heavily skewed. Examination of a density plot will reveal whether the empirical distribution is heavily skewed. Figure 6.5 contains a density plot for the relationship duration variable. The variable is, as expected, heavily skewed with a long tail for positive support. Therefore, we might use a quadratic transformation to model a threshold, or we might use a log transformation given the skew. With standard modeling techniques, there isn't any way to discriminate between

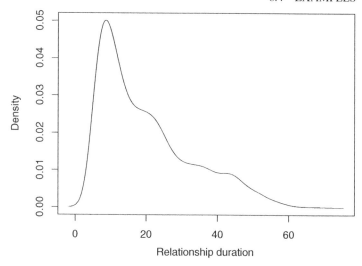

Figure 6.5   Density plot for relationship duration.

these two transformations. A GAM, however, allows us to test these transformations against the spline fit. We tested the GAM against both transformations. We find that the logarithmic transformation is inadequate as the likelihood ratio test indicates that the semiparametric model provides a significantly better fit ($p = 0.029$), but the difference in fit between the model with a quadratic term and the semiparametric model is not statistically significant ($p = 0.16$). The test indicates that a quadratic transformation is an adequate model for the nonlinearity, and the analyst could comfortably replace the spline fit with a quadratic transformation. The purpose of this illustration has been to demonstrate that even if one does not settle on a semiparametric specification, the GAM is useful for testing between power transformations.

Finally, we should note that in this instance there is no obvious statistical reason for using the quadratic transformation over the spline fit. One could argue for the transformation on grounds of parsimony. But the estimate in Figure 6.3 required three degrees of freedom for the smoothing spline fit, so using the quadratic transformation, we save one degree of freedom. In a model with over 4000 cases, the estimation of one extra parameter matters little. If degrees of freedom are plentiful, analysts can choose either the transformation or nonparametric estimate if the fit is no different. In this instance, however, the calculation of predicted probabilities is less complex with a quadratic term, so one might opt to forgo the semiparametric model.

6.4.3   Count Models: Supreme Court Overrides

We now turn to an illustration of how to estimate count models such as Poisson and negative binomial regression models as GAMs. For the most commonly used models, the stochastic component is assumed to follow either a Poisson or negative binomial distribution with an exponential link function. For readers not familiar with these models both Long (1997) and Cameron and Trivedi (1998) are useful references. Again, the systematic component of the model can be linear and additive or the analyst can use nonparametric techniques to allow for more complex functional forms. The point of this example, however, is not to demonstrate that it is possible to estimate a GAM for count data. Instead, we demonstrate how specification errors caused by undiagnosed nonlinearity can infect other terms in the model and lead to misleading diagnostics for count models. While misspecification due to incorrect functional form can occur in any GLM, for count models it can influence the actual choice of the model as we will see.

For this illustration, we analyze data on the number of Congressional acts overturned by the Supreme Court each year.[5] The data are on the First through the 101st Congress (1789–1990). The number of acts overturned each year is bounded below by 0, is discrete, and is typically a small number each year, which suggests using a Poisson distribution instead of a Normal for the stochastic component of the GLM. Dahl's (1957) study of the US Supreme Court as a countermajoritarian institution suggests several relevant predictors of the number of Congressional acts overturned by the Supreme Court. He suggests that when Congressional unity is high the Court will be less likely to overturn laws and that justices with more experience will be more likely to challenge the authority of Congress.

We model the number of overturned acts as a function of a dummy variable that indicates if both chambers of Congress are controlled by the same party and a measure that records the average number of years served by the justices sitting on the Court that year. We also include a counter for the Congress to control for the institutionalization of judicial review and to capture the general propensity of the Court to overturn Congress. Given that the time span of the data is over 200 years, we should expect the Court's willingness to overturn Congressional acts has varied over this period. Including the counter on the right hand side of the model as linear term would clearly be the wrong functional form, since that would imply that we believe the willingness of the Court has increased (or perhaps decreased) every year since 1789. A quadratic transformation would also probably be incorrect since it would imply that there has been an increase in the number of acts overturned followed by a decline over the period. More likely is

---

[5]The data and the logic of the basic analysis were generously provided by Chris Zorn.

that propensity of the Court to challenge Congress has probably increased and decreased several times over this period. While we expect that the Court's willingness to confront Congress has varied, we do not have any *a priori* knowledge of what the functional form for this temporal variation might be. Transformations are, of course, one possibility, but it makes little sense to assume that the temporal dependence is quadratic or cubic, when a spline fit will estimate the nonlinearity from the data. Automated smoothing techniques are a good choice with such terms since the form of the nonlinearity may be complex. For this illustration, we estimated a parametric Poisson regression model and a semiparametric Poisson model with smoothing splines and GCV selected smoothing. The results for both models are in Table 6.1.

Careful examination of Table 6.1 shows that the estimated parameter for the tenure variable is almost 2.5 times larger in the GAM than for the parametric model. This is a good example of how failure to diagnosis nonlinearity in the model specification can have implications for other estimates in the model. This is particularly true for GLMs where the variables need not be correlated for misspecification to cause biased estimates. To further demonstrate the bias caused by the failure to model nonlinearity, we plot the expected count of Supreme Court

Table 6.1 A comparison of parametric and semiparametric models of Supreme Court overrides.

	Supreme Court overrides (parametric)	Supreme Court overrides (semiparametric)
Justice	0.07^{*}	0.19^{***}
Tenure	(0.03)	(0.05)
Unified	0.14	0.18
Congress	(0.24)	(0.28)
Congress Counter	0.03^{***}	$-^{***}$
	(0.003)	
Constant	-2.7^{***}	-3.31^{***}
	(0.52)	(0.77)
Deviance explained	44%	67%
LR Test p-value		0.00

Likelihood ratio test against previous model in the table.

Standard errors in parentheses. Two-tailed tests.

$^{*}p$-value < 0.05 ** p-value < 0.01 *** p-value < 0.001

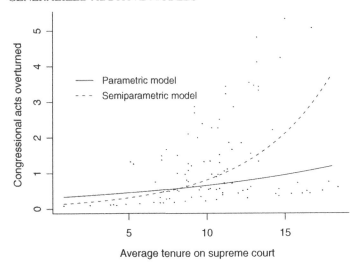

Figure 6.6   The difference in the effect of tenure across parametric and semi-parametric models.

overrides across the range of the tenure variable from the two models. Figure 6.6 contains a plot of the effect of tenure on the expected count of Supreme Court overrides for both models. The attenuation of the effect from the parametric model is quite clear in the plot.[6] The use of the semiparametric model has a large substantive effect on the inferences we make with this model. The specification error most likely stems from overestimating the over time change in the Court's willingness to confront Congress.

As we might expect, the Poisson regression model with smoothing applied to the Congress counter provides a better fit to the data. The evidence from a likelihood ratio test indicates that the effect of the change in Congress over time is nonlinear as the χ^2 test statistic of 50.79 on 7.6 degrees of freedom is highly significant ($p < 0.001$). Moreover, the deviance explained improves from 44% to 67%. It is possible that some transformation of the counter variable will capture the temporal dependence. We compared a model with quadratic and cubic terms against the GAM. Again, the GAM fit is superior as the χ^2 test statistic is 31.99 on 5.6 degrees of freedom ($p < 0.001$) for the comparison to the cubic specification. Unlike many of the examples in this book, the nonlinearity in this model takes a more complex form. Figure 6.7 contains the a plot of the nonparametric estimate for the Congress counter variable. The spline estimate in the plot

---

[6]I omit confidence bands for the expected count plots to avoid cluttering the figure.

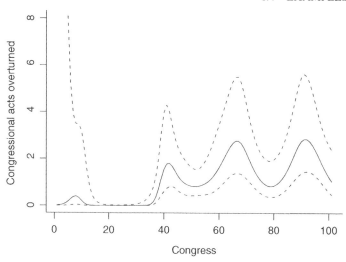

Figure 6.7   Nonparametric estimate for Congressional counter variable.

clearly demonstrates that the Court's propensity to overturn Congressional acts has varied widely over time, with the Court being increasingly activist over time but with considerable variance.

Poisson regression models are often deficient since the variance of the count exceeds the mean, leading to overdispersed counts. When overdispersion is present in the data, a negative binomial model, which accounts for the difference between the mean and variance in the count, is used instead. One method for detecting overdispersion is to compare the negative binomial model against the Poisson model using a likelihood ratio test. But as Cameron and Trivedi (1998) note, overdispersion may signal basic misspecification of the mean for the Poisson regression model. They caution against automatic use of the negative binomial model without further consideration of the Poisson specification. An example of this arises with the data in this illustration. The mean and standard deviation for the overrides variable are 1.24 and 1.79 respectively, so the variance exceeds the mean by a modest amount. If we test for overdispersion with the counter variable included as a linear term, we find that the negative binomial model is a better fit to the data ($p = 0.004$). If we test for overdispersion in the Poisson GAM, however, we find that the negative binomial model does not fit the data any better ($p = 0.147$). Correctly modeling the functional form for the counter variable improves the Poisson specification to the point that overdispersion is no longer problematic.

6.4.4   Survival Models: Race Riots

Nonparametric estimates can be incorporated into the right hand side of event history models as well. We assume that readers have some basic familiarity with survival models and the Cox model in particular. For readers not familiar with these models, Box-Steffensmeier and Jones (2004) provide a good introduction. For parametric event history models, such as the log-logistic, splines are a means of modeling nonlinearity. But controlling for nonlinearity is particularly important in the Cox proportional hazards model where nonlinear effects can be mistaken for nonproportional hazards. In this example, we review the concept of nonproportional hazards and then demonstrate the use of splines with a Cox model with a data set for the onset of racial rioting. A key assumption for many survival models is that the hazard ratios are proportional to one another and that proportionality is maintained over time. This is true for the widely used Cox model. Box-Steffensmeier and Jones (2004) review how the assumption of proportional hazards may be violated with social science data, and they outline the diagnosis of and corrections for nonproportional hazards. Correcting for nonproportional hazards is critical since it leads to biased parameter estimates and reduced statistical power.

Violation of the nonproportional hazards assumption is typically caused by a time-varying covariate effect. Diagnostics for nonproportional hazards, such as Schoenfeld residual tests, detect both nonproportionality and nonlinear function forms (Therneau and Grambsch 2000). Correcting for nonproportional hazards with log-time interactions when the model failure is due to incorrect functional form will not rectify the misspecification. The relationship between nonproportional hazards and the correct functional form for covariates suggests a diagnostic sequence for the Cox model. First, the correct functional form for the covariates needs to be found and fitted before testing for nonproportional hazards. Proceeding in this order will ensure that signs of nonproportional hazards are not due to nonlinearity.

In this example, we use data from Myers (1997) to demonstrate the use of spline fits in a Cox model and show how undiagnosed nonlinearity can appear as nonproportional hazards. Myers analyses how various local conditions led to the onset of racial riots in 410 US cities from 1961 to 1968. He investigates how rioting spread through diffusion while controlling for local conditions. In our analysis, we replicate one of his models where the onset of riots was a function of the number of non-white unemployed (in 1000s), the median manufacturing wage, the unemployment rate, the percentage of foreign born residents, an interaction between non-white and unemployment, a measure of spatial diffusion, a measure of national-level diffusion, a dummy for the years 1967 and

Table 6.2 Grambsch and Therneau nonproportionality tests for time to onset of racial riots, 410 US Cities, 1961–1968.

	Nonproportionality test p-value (no corrections)	Nonproportionality test p-value (Cox GAM)
Number of non-white unemployed	0.13	0.90
Median manufacturing wage	0.69	0.81
Unemployment rate	0.42	0.28
Percent foreign-born	0.47	0.68
non-white unemployed × percent foreign-born	0.07	0.84
Spatial diffusion	0.57	0.83
National level diffusion	0.95	0.67
National level diffusion squared	0.61	—
Prior rioting	0.25	0.40
Global test	0.09	0.99

1968, and a measure of prior rioting. In the original analysis, the measure for non-white unemployment was logged and the measure of national-level diffusion was squared.

We estimated a replication model and a Cox GAM using smoothing splines for all the continuous predictor variables. Then using both a visual inspection of the nonparametric estimates and a series of likelihood ratio tests, we found that while the unemployment rate and percent foreign born had linear effects, the rest of the terms had nonlinear effects.[7] We then conducted Grambsch and Therneau nonproportionality tests for both the replicated model and the final Cox GAM specification. The results are in Table 6.2.

In the replication model, nonproportional hazards are not widespread, but the global test p-value is close to the standard 0.05 threshold as is the p-value for the interaction. The results for the GAM are much improved as the global test is no longer borderline. Moreover, several of the p-values are much larger for the test under the GAM. As a general rule, the inclusion of nonlinear effects

---

[7]We did not model the interaction as a nonparametric interaction, as currently no software exists for such an analysis with a Cox model.

will improve the test results for nonproportional hazards. If signs of nonproportional hazards remain, the analyst should proceed to use the strategies for correcting nonproportionality. The reader should be aware that when testing for nonproportional hazards using the Grambsch and Therneau test, the software is programmed to return a test statistic and p-value for the interval between each knot. This may mean 10–12 test statistic values for each covariate. One must look at each one. Here, we reported an average p-value across all the values returned. See Keele (2006) for other examples of how undiagnosed nonlinearity can appear as nonproportional hazards.

Finally, Figure 6.8 contains plots for two of the nonparametric terms in the model. For event history models, the nonparametric effect is plotted against the log hazard for the model. A rising hazard implies that the event of interest is more likely to happen, while a falling hazard implies that the event is less likely to happen. In this example, the event is the onset of a racial riot in the 1960s, so a rising hazard implies that a riot is more likely to occur. The first panel of Figure 6.8 contains a plot of the nonparametric estimate for manufacturing wages. The hazard of a riot rises, is flat, and then declines sharply. The second panel contains the nonparametric estimate for non-white unemployment. The effect is interesting as it is appears to be a double threshold. Here, we see that the hazard of a riot increases strongly until unemployment is approximately 5% and then it levels off, but once unemployment rises above approximately 10%, the hazard of a riot again increases before leveling off a second time. The use of a GAM also changes the inferences we draw from the model. In the original analysis, the interaction between non-white unemployment and percent-foreign born was not statistically significant, but once nonlinearity is accounted for, it becomes highly significant.

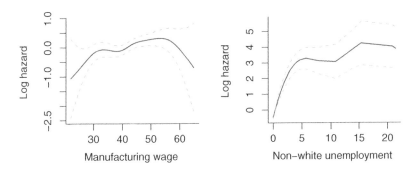

Figure 6.8   The effect of manufacturing wages and non-white unemployment on the hazard of racial riots, 1961–1968.

Like for other GLMs, the addition of nonparametric terms to event history models allows for diagnosing and modeling nonlinearity. Correct diagnosis of nonlinearity is doubly critical for Cox models, however, since nonlinearity can appear as nonproportional hazards in diagnostic tests.

6.5 Discussion

In this chapter, we extended the semiparametric regression model to generalized linear models. GLMs retain a key linearity assumption that if incorrect leads to specification errors. Semiparametric techniques allow an analyst to diagnose and model nonlinearity if it is present. Again all the lessons learned from smoothing and the semiparametric regression model apply to GAMs. Just as in Chapter 5, one needs to take care to prevent overfitting, and if using automatic smoothing techniques, the p-values for the likelihood ratio test are again approximate. Finally, nonparametric estimates ask more of the data than parametric models. At times, GAM will fail to converge where convergence is not a problem when the model is estimated as a GLM. Such failures of convergence are not common but can occur. When convergence fails, one must rely on fully parametric models.

6.6 Exercises

Data sets for exercises may be found at the following website: http://www.wiley.com/go/keele_semiparametric.

1. Use the data set `turnout.dta` which contains measures of age, education in years, region as indicated by an indicator variable for those respondents who live in the South, an indicator for whether the election in question was for governor, and the days between election day and the date voter registration closed and whether the respondent voted in the 1988 US presidential election. This data set is a subset of the full Current Population Survey data that Nagler (1991) analyzed.

   (a) Estimate a GAM with nonparametric terms for the three continuous covariates. Plot the estimates. Do any appear to be linear or nonlinear? Check your conclusions with a series of likelihood ratio tests.

   (b) For any covariates that appear to have a nonlinear effect, test the spline fits against quadratic and logged fits. Which model do you prefer?

   (c) Reestimate the model using unpenalized splines. Repeat the likelihood ratio tests. Do the p-values from the tests differ when compared to those for the models fit with smoothing splines and automated smoothing? Why might we expect them to differ? Which set of *p*-values are the most accurate?

(d) Plot the nonparametric estimate of age in both the linear scale and on a probability scale.

(e) Finally, fit a nonlinear interaction between age and education. Plot the estimate. Does it appear to be nonlinear in both dimensions? Use a likelihood ratio test to see if the nonlinear interaction should remain in the model.

2. The data set `drvisit.dta` contains data on the number of doctor visits along with a series of predictor variables. The data for this exercise have been analyzed by both Cameron and Trivedi (2005) and Deb and Trivedi (2002) using parametric models. There is also a data description file available.

(a) Fit a semiparametric Poisson regression model using nonparametric estimates for all the continuous variables. Plot the results and then use likelihood ratio tests to determine which variables require nonparametric fits.

(b) For those variables that appear to be nonlinear, test whether power transformations model the nonlinear adequately.

(c) Decide on a final specification for the model. Estimate this model and provide an interpretation of what you find.

(d) Plot any nonparametric estimates from the model on the scale of expected counts.

(e) Now estimate the model as a negative binomial regression model and plot the nonparametric estimates. Is there any difference between the plots?

For the next example you will use the same data from the logistic regression example in this chapter. Logit models with smoothed time terms are equivalent to the Cox proportional hazards model. Use the data set `PHdata.dta`. There is also a command file that replicates the logit model and demonstrates how to estimate the equivalent Cox model with and without nonparametric terms.

(a) Fit a semiparametric regression model to the data using spline fits for all the continuous predictors. Plot all the nonparametric estimates. Which appear to be nonlinear?

(b) Now use a series of likelihood ratio tests to decide which variables have nonlinear effects.

(c) Test for nonproportional hazards with and without nonparametric terms. Do the test results differ?

(d) Decide on a final specification for the model. Use either time or log time interactions for variables that have nonproportional effects.

7

Extensions of the Semiparametric Regression Model

Thus far we have used nonparametric smoothers to estimate semiparametric versions of familiar models such as linear and logistic regression. In this chapter, we take up three different applications of nonparametric regression models. First, we describe using smoothers to estimate semiparametric mixed models. Here, we use smoothers to test for and model nonlinearity in the usual way. The extension of smoothers to mixed models is fairly straightforward but can produce estimation difficulties. Next, we provide an overview of using Bayesian estimation techniques such as Markov Chain Monte Carlo (MCMC) to estimate semiparametric regression models. Using MCMC to estimates splines and semiparametric models is not overly complex and mostly an exercise in programming, but it does introduce a few complications not present when estimating semiparametric models with standard techniques. Finally, I consider the use of GAMs for estimating propensity scores. The estimation of propensity scores is a natural application of GAMs, but the nature of propensity scores requires care in evaluating whether the GAM is appropriate.

Semiparametric Regression for the Social Sciences   Luke Keele
© 2008 John Wiley & Sons, Ltd

7.1   Mixed Models

In recent years, mixed models, known as multilevel models or hierarchical linear models, have seen increased use in the social sciences.[1] Researchers in the social sciences use these models to understand how context affects individual level behavior, but mixed models are useful any time we have sets of observations that fall into groups that share common characteristics. When analysts estimate mixed models, they typically assume linear functional forms. As with other parametric models, we have no *a priori* reason to assume that the linear functional form is correct for a given mixed model. Again, while a linear functional form might be appropriate, we can use nonparametric regression models to both diagnose and model nonlinearity in mixed models. In this section, we briefly outline linear mixed models and demonstrate how semiparametric techniques can work within such models. We end with an application to a well-known data set on educational outcomes.

Linear Mixed Models

The outline of mixed models that follows is best read as review for the reader already familiar with such models. For readers unfamiliar with such models, there are number of excellent texts on this topic such as Raudenbush and Bryk (2002) and Gelman and Hill (2006). Mixed models in the social sciences typically have levels with different equations at each level. While three and four levels models are possible, models with more than two or three levels are rare. We begin with the level-1 model:

$$Y_{ij} = \beta_{0j} + \beta_{1j}X_{ij} + e_{ij} \tag{7.1}$$

where we have $i = 1, \ldots, N$ observations nested in $j = 1, \ldots, J$ units. The level one observations might be nested in states, electoral units, counties, or schools. The parameters β_{0j} and β_{1j} are thought to vary across the j units. The variation in the β parameters is represented with the following level-2 equations:

$$\beta_{0j} = \gamma_{00} + \gamma_{01}Z_j + u_{0j}$$
$$\beta_{1j} = \gamma_{10} + \gamma_{11}Z_j + u_{1j}. \tag{7.2}$$

These equations imply that the slope and intercept varies systematically as a function of group level characteristics and random variation as captured by u_{0j}

---

[1]Multilevel models are actually a special case of mixed models. The mixed model takes a wide variety of forms with the multilevel being just one of those forms. In this chapter, however, we use the terms interchangeably.

and u_{1j}, the random effects. Here Z_j represents a variable measured at the group level or level-2 which is thought to predict the variation in the level-1 parameters. Substituting the level-2 equations into the level-1 equation produces the following single equation:

$$Y_{ij} = \gamma_{00} + \gamma_{01}Z_j + u_{0j} + (\gamma_{10} + \gamma_{11}Z_j + u_{1j})X_{ij} + e_{ij}$$
$$= \gamma_{00} + \gamma_{01}Z_j + \gamma_{10}X_{ij} + \gamma_{11}Z_jX_{ij} + u_{1j}X_{ij} + u_{0j} + e_{ij}. \quad (7.3)$$

Each term in the above equation has a specific interpretation. The parameter γ_{00} is referred to as the grand mean and represents the average of Y_{ij} across all the j units. The parameter γ_{01} represents deviations from the grand mean as predicted by changes in Z_j. The term γ_{10} is the average slope or the effect of X_{ij} on Y_{ij} across the j units. The parameter γ_{11} represents deviations from γ_{10} across the values of Z_j. The term u_{0j} is a random effect at level-2 across the j units; this term represents random factors that cause units to deviate from the grand mean. Finally, u_{1j} is the random effect of unit j on γ_{10} controlling for Z_j.

The random effects, u_{0j} and u_{1j}, and the level-1 error, e_{ij} are assumed to be drawn from a normal distribution. The level-1 error variance σ^2 measures the variance in Y_{ij} at level-1. The term σ^2 is called the the within-group variability, as it measures how much observations within the same unit vary from the unit mean. The variances for the random effects are τ_{00} and τ_{11} respectively. The variances for the random effects estimate the amount of variability in the intercepts and slopes between groups. In other words, these terms estimate how much each observation varies from the grand mean and average slope across all units.

Generalized Additive Mixed Models

The semiparametric framework can also be extended to mixed models. Nothing in the development of the mixed model precludes a nonlinear functional form, and we can extend the semiparametric framework to mixed models to test for and model nonlinearity. The resulting model is the Generalized Additive Mixed Model (GAMM). GAMMs incorporate splines into mixed models to allow the analyst to test for and model nonlinearity. We rewrite the mixed model from the previous section as a GAMM in the following equation:

$$Y_{ij} = \gamma_{00} + \gamma_{01}Z_j + f_1(X_{ij}) + f_2(Z_jX_{ij}) + u \quad (7.4)$$

where u is a compound error term comprised of $u_{1j}X_{ij} + u_{0j} + e_{ij}$. The above model relaxes the assumption that the effect of X_{ij} is linear but retains the mixed model structure where the effect of X_{ij} differs across the levels of Z_j.

If we estimate such a model, the estimate of f_2 takes the form of the interaction effects estimated in Chapter 5. If Z_j is discrete, we plot the estimate of f_2 for each value of Z_j. If Z_j is continuous, the estimate of will f_2 be a three-dimensional plot of how the effect of X_{ij} depends upon Z_j. The analyst can use likelihood ratio tests to decide whether the nonparametric term or terms are necessary.

At this point, an example is useful. The data for this example are from the 1982 High School and Beyond survey of 7185 high school students in 160 different schools. These data identical to those used in the early chapters of Raudenbush and Bryk (2002). The response variable is math achievement scores which are thought to be a function of the student's family socioeconomic status (SES) and whether the school is a parochial or public. For this data, we might write the following mixed model

$$Y_{ij} = \gamma_{00} + \gamma_{01} Z_j + \gamma_{10} X_{ij} + \gamma_{11} Z_j X_{ij} + u \qquad (7.5)$$

where Y_{ij} is a student's math achievement score, X_{ij} is the student's family SES, and Z_j is dummy variable which indicates whether the school is parochial or not, and $u = u_{0j} + e_{ij}$ which implies that there is no random effect across the level-1 slope coefficients.[2] Importantly, X_{ij} is a student level variable, while Z_j is a school level variable and γ_{11} captures the effect of school type on student SES, which implies that the effect of SES on math achievement differs across parochial and public schools.

If we estimate this mixed model, we find that $\hat{\gamma}_{00} = 11.41$, $\hat{\gamma}_{01} = 2.80$, $\hat{\gamma}_{10} = 2.78$, and $\hat{\gamma}_{11} = -1.34$. These estimates imply that the average level of math achievement in public schools is 11.41, and that the average level of achievement in Catholic schools is 2.80 points higher. A one point increase in SES increases math achievement 2.78 points in public schools, but that effect is -1.34 points weaker in parochial schools. Thus we see that average achievement is higher in parochial schools, and there is a weaker relationship between SES and achievement in those same schools. We might suspect, however, that the effect of SES is nonlinear due to diminishing returns, but given that the relationship between SES and achievement is weaker in parochial schools, we should still allow the effect SES to vary across school type. To allow SES to have a nonlinear effect that varies across school type, we estimate the following GAMM

$$Y_{ij} = \gamma_{00} + \gamma_{01} Z_j + f_{10} X_{ij} + f_{11} Z_j X_{ij} + u. \qquad (7.6)$$

---

[2]The decision to omit this random effect on the coefficient is done in light of prior analyses we have conducted with these data.

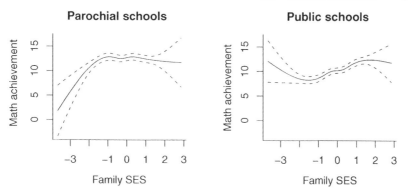

Figure 7.1   The effect of family SES on math achievement for parochial and public schools.

For the above model, the nonparametric estimates will be comprised of two plots, one for the effect of SES on achievement in Catholic schools and another for the effect of SES on achievement in public schools. Figure 7.1 contains a plot of the nonparametric estimates for the above model. The first panel of Figure 7.1 contains an estimate for SES in parochial schools. The effect is clearly nonlinear as math achievement is strongly related to SES until SES reaches average levels, but once SES reaches higher levels, it has little affect on test scores. For public schools, the effect is nearly linear if one ignores the end points where the confidence intervals spread out. The use of a GAMM is informative. Using a standard mixed model, we conclude that the effect of SES on achievement is weaker in parochial schools than in public schools. The GAMM reveals, however, that the effect differs in an important way. For parochial schools, SES only matters at the bottom end of the scale, but for public schools there is a more uniform effect. A likelihood ratio test indicates that the GAMM is a better fit to the data than a parametric mixed model ($p < 0.001$).

In sum, semiparametric techniques readily extend to mixed models. Given the prevalence of interactions in mixed models of this type, nonparametric regression techniques provide considerable flexibility to capture nonlinear interactions between level-1 and level-2 variables. Unfortunately, this extension of semiparametric regression cannot be done without some cost. The mixed model is a more computationally intensive model to fit than the standard linear model, and a GAMM adds to the computational burden. Typically, the only noticeable difference is a slightly increased estimation time. However, particularly if the level-1 model is a GLM, estimation is more difficult, and the model may fail to converge. If this occurs one must estimate a simpler model.

7.2   Bayesian Smoothing

Traditionally, estimation and inference in the social sciences is done within the frequentist framework of probability. Whether one uses OLS or a maximum likelihood estimator, inference is based on the principles of repeated sampling and parameters estimated through maximization.[3] While the frequentist paradigm is still dominant in the social sciences, in recent years the Bayesian paradigm has made considerable in-roads. Nonparametric and semiparametric regression models can be estimated within the Bayesian framework. Below, we briefly compare the frequentist and Bayesian paradigms, but the comparison that follows is superficial at best. For more detailed introductions to Bayesian inference see Jackman (2000), Gill (2002) and Gelman *et al.* (2003). We then translate semiparametric regression models into the Bayesian paradigm with a discussion of the appropriate priors, basis functions, and implementation. We conclude with an example of a Bayesian semiparametric regression model.

The Frequentist and Bayesian Approaches

Under the frequentist approach to statistical models, the parameter to be estimated, θ, is an unknown but fixed quantity. A random sample is drawn from the population of interest, and based on the sample drawn, estimates of θ and the precision of that estimate are obtained. Estimation of the parameter is typically done through either maximum likelihood estimation or ordinary least squares. In the Bayesian framework, θ is not assumed to be fixed but instead is a quantity that varies and is described by a probability distribution. The probability distribution that characterizes θ is called the *prior distribution*. The prior distribution for θ is formulated by the analyst before estimation. A random sample of data is used to update the prior distribution through Bayes' rule. The updated prior is called a posterior distribution, and the posterior distribution is used to make inferential statements about θ, which remains a random quantity. For example, the mean of the posterior distribution is often used as an estimate for θ (Casella and Berger 2002). Typically, analytic derivations of the posterior distribution are intractable, and the posterior distribution is estimated through Markov Chain Monte Carlo (MCMC) simulation.

Are there any reasons to prefer Bayesian estimation for nonparametric smooths and semiparametric regression models? Some analysts prefer Bayesian estimation for any model on philosophical grounds. They argue that Bayesian estimation is a

---

[3]While OLS is by definition minimizing the sum of squares, it can be easily demonstrated that this is a maximization problem.

more realistic account of probability. Specifically, Bayesians argue that given that we never know θ, it makes more sense to think of it as a random quantity. There are also a few practical reasons one might prefer a Bayesian framework. The MCMC algorithms used in Bayesian estimation are very robust and can produce estimates for complicated models. Given that semiparametric regression techniques adds a computational burden to the estimation, this can be useful. Finally, Bayesian inference provides us with the confidence intervals that fully reflect our uncertainty about the amount of smoothing applied to the data. Automatic smoothing techniques make it difficult for us to fully capture the estimation uncertainty for the nonparametric estimate. In the Bayesian model, the percentiles of the posterior distribution serve as the confidence intervals, and these confidence bands fully reflect our uncertainty about the smoothing estimate.

Bayesian Semiparametric Regression

There are several different methods for estimating regression models that include splines in a Bayesian framework. We outline the approach of Ruppert, and Wand (2005). For an alternative approach, see Congdon (2003). In many ways, Bayesian semiparametric models differ little from those we have estimated thus far. We will still use splines to estimate nonlinear effects for continuous covariates, and we can still estimate some of the terms in the model parametrically, and Bayesian estimation can be used for semiparametric regression models, GAMs, and GAMMs. We will rely on the mixed model representation of splines from Chapter 3 as the basis for the estimation of splines in a Bayesian context. Recall that we could write smoothing splines as a mixed model

$$\mathbf{Y} = \mathbf{X}\boldsymbol{\beta} + \mathbf{Z}\mathbf{u} + \varepsilon \tag{7.7}$$

where $\mathbf{X}$ is a matrix that contains a constant and a single exogenous variable, $\boldsymbol{\beta}$ is a vector of regression coefficients, $\mathbf{Z}$ is a matrix of the basis functions that define the knot locations, and $\mathbf{u}$ is a vector of random effects placed at each of the knot locations. The mixed model representation is useful for Bayesian estimation, since it provides a statement of the likelihood for the spline model. The posterior distribution is proportional to the product of the prior distribution and the likelihood, and thus we require a likelihood for the spline model. For Bayesian estimates, we must place prior distributions on the model parameters and use MCMC to simulate the posterior distribution.

There is, however, one added complication. Before, the choice of basis functions was not of great importance, and we relied upon cubic basis functions. For Bayesian estimates, the choice of the basis function has important consequences for the properties of the MCMC sampling. For Bayesian estimates, low-rank thin plate splines are the best choice for the basis functions, since they improve the mixing of the MCMC chains. The posterior correlations between the parameters are much smaller when thin plate basis functions are used, which greatly improves the mixing of the chains (Crainiceanu, Ruppert, and Wand 2005). See Chapter 3 for details on the thin plate basis functions.

Thus far, we have developed a Bayesian model for nonparametric smoothing but not for a semiparametric regression model. The move to a Bayesian semiparametric regression model is not difficult. Additional covariates that are to have a parametric representation are simply added to the **X** matrix as additional fixed effects. The Bayesian semiparametric regression model can be estimated using the Gibbs sampler in WinBUGS. One advantage of the Bayesian semiparametric regression model is that the programming for the model is transparent. With the exception of the code for the Gibbs sampler, the basis functions and the model structure are readily apparent to the analyst and easily altered. This provides greater flexibility than with standard estimation techniques for semiparametric models which are not easily altered.

We estimate a Bayesian semiparametric regression model for the model of challenger's vote share from Chapter 5. Recall that in this model the challenger's vote share was a function of challenger spending, incumbent spending, presidential vote share, a dummy variable for an experienced challenger, a dummy variable for whether the district was competitive in the last election, a dummy variable for whether the district had been redistricted, the number of overdrafts the incumbent incurred against the House Bank, and the support for Perot in the district. We estimate a simplified model that includes the measures of challenger spending, presidential vote, the number of overdrafts, and the dummy variable for redistricting. In particular, we focus on the nonlinear relationship between challenger spending and the challengers' vote share controlling for the other variables. Before estimating the model, we select prior distributions for the model parameters. We use the following prior distributions:

$$\begin{cases} \beta & \sim N(0, 10^6) \\ \sigma_\beta^{-2}, \sigma_\varepsilon^{-2} & \sim \text{Gamma}(10^{-6}, 10^{-6}). \end{cases} \tag{7.8}$$

Both are uninformative priors. The prior for β is Normally distributed, and we use the Gamma(a,b) distribution for the variance terms so that the mean is $a/b = 1$ and the variance is $a/b^2 = 10^6$. Crainiceanu, Ruppert, and Wand (2005) provide a more detailed discussion of the choice of the priors for the variance terms, since the Gamma prior may not be sufficiently uninformative depending upon the scaling of the variables in question. Gelman *et al.* (2003) and Natarajan and Kass (2000) outline alternatives to the Gamma priors.

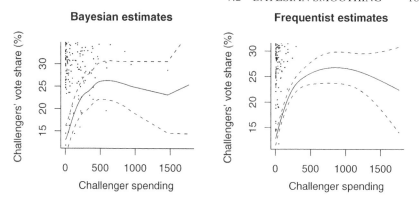

Figure 7.2   Comparison of Bayesian and frequentist estimates of challenger spending on challenger's vote share.

Estimating this model in **WinBUGS** requires the analyst to manually select the number of knots and their location. Since we are using smoothing splines for the nonparametric estimates, we should use a large number of knots placed evenly along the range of the variable to be smoothed. For this model, we use 20 knots placed at equal intervals in the range of the challenger spending variable. We estimated the model with different knot numbers, but the estimated were unchanged. To estimate this model we used 20 000 MCMC iterations for a single chain with a burnin of 5000 iterations. Model convergence was apparent in the traceplots after 5000 iterations.

Figure 7.2 contains two nonparametric estimates of challenger spending on challenger vote share. The first panel contains the Bayesian estimate from **WinBUGS**. The second panel contains the estimate from a penalized IRLS algorithm. The estimates are similar and lead to the same basic conclusion about the dependency between challengers' vote share and spending. These fully Bayesian confidence intervals completely capture our uncertainty about the smoothing parameter estimate while the frequentist confidence bands do not. The Bayesian confidence intervals do appear to be slightly wider than the frequentist confidence bands. However, one would be hard pressed to say that it makes much difference.

At this point, one might wonder if using the Bayesian framework exclusively might be recommended. We do not recommend such an approach. For many simple models where one wants to test a few continuous covariates for nonlinear effects, the standard nonparametric models are far more convenient. Moreover, for many standard models, the added time required for the programming is not worth the minor differences found with Bayesian estimates. Where the Bayesian framework shines is the flexibility that it provides to the analyst. In the standard framework, one is often at the mercy of the software for which model smoothing can be added to, but the Bayesian model allows the analyst to incorporate smoothing to any model in a straightforward manner. The added computational cost of smoothing can make it difficult to estimate the desired model, and the robustness of MCMC simulation can be useful in such situations as it can produce estimates for models that are computationally intensive.

7.3   Propensity Score Matching

Statistical analysis may be used for both data reduction and causal inference. When interested in causal inference, the analyst attempts to estimate whether a change in conditions or circumstances leads to an outcome. In the analysis of data, the analyst can most safely make causal inferences when the study is based on a randomized experiment. In randomized experiments, the investigator randomly assigns subjects to either a treatment or control group and then administers a stimulus to the treatment group. Random assignment of the subjects balances the two groups, up to random error, with respect to all relevant factors other than the treatment. Any differences found between the control and treatment groups are attributed to the treatment.

Of course, experiments cannot be used in many circumstances. They are often expensive and impractical for some areas of study. Moreover, in many situations, the use of an experiment would be unethical.[4] When experiments are infeasible, we often collect and analyze observational data. In an observational study, we can only observe the outcome after the subjects have assigned themselves to groups. The analysts can still compare group differences, but any differences that are observed may be due to some third confounding variable that is associated with both the treatment and the outcome. In observational studies with regression methods, the problem of confounders is dealt with by introducing control variables. The use of control variables retains strong functional form assumptions that may be unrealistic even with the introduction of nonlinear terms (Winship and Morgan 1999). Moreover, regression based method tend to obscure absences of common support.

An alternative to regression based methods is matching. Matching is a technique designed to bring the logic of experiments to observational data (Rosenbaum and Rubin 1983a,b, 1984, 1985a,b). To understand matching, some notation is helpful. Let Y_i be some outcome for each individual i. We also observe

---

[4]In a study of smoking, for example, it would be unethical to randomly assign some subjects to smoke when it is suspected of being an addictive health risk.

T_i an indicator of whether each individual received the treatment or not and $\mathbf{X}_i$ is a set of observed characteristics for each observation. Under matching, we attempt to match each observation in the treatment group with an observation (or observations) in the control group that is identical in all respects on the factors measured in $\mathbf{X}_i$.

Matching has several advantages over the traditional regression framework. Under matching, no assumption is made about the functional form of the relationship between Y_i and $\mathbf{X}_i$. Second, only the observations that are in either the control or treatment group enter into the estimation of the treatment effect ensuring that inferences are based on common support. Matching also allows researchers to make study design choices without observing how those choices affect the study outcome. In a regression based design, analysts can cull specifications in search of the 'best' fit based on statistical significance. The ability to make research designs decisions without knowing how each decision affects the conclusions of the study is thought to contribute to greater honesty in studies of observational data (Rubin 2001). Finally, matching requires fewer parameters, so it is more efficient than regression designs based on similar specifications. The drawback is that matching suffers from a serious dimensionality problem. Unless one has a very large data set, it becomes impossible to match on more than a few X_i's. The propensity score solves this dimensionality problem.

Propensity Scores

Rosenbaum and Rubin (1983b) found that an alternative way to condition on $\mathbf{X}_i$ is to match on the probability of being assigned to the treatment often called the propensity score. The propensity score has the great advantage of being a one-dimensional metric that we can match on. Rosenbaum and Rubin (1983b) prove that the propensity score contains all the relevant information from $\mathbf{X}_i$ required for a balanced model. A balanced model is one where the treatment and control groups do not differ with respect to $\mathbf{X}_i$ in any way that is related to the treatment. Propensity scores are also easy to estimate. For the propensity score, we need an estimate of $\Pr(T = 1|\mathbf{X}_i)$. This can be quite readily obtained with logistic regression. The indicator of treatment is regressed on $\mathbf{X}_i$ and the resulting probabilities are used for matching. Once matching is completed, the analyst must check that treatment and control groups are balanced. That is, he or she must compare the matched treatment and control groups to ensure that the distributions of the two groups with respect to $\mathbf{X}_i$ are the same.

To improve balance often requires respecification of the propensity score model to include interactions and power transformations. One might assume that a GAM might be a natural choice for the estimation of propensity scores given that it is standard practice to use power transformations on continuous covariates in $\mathbf{X}_i$. The use of a smoothed terms can be a useful method for improving balance, but the application of GAMs to the estimation of propensity scores is not entirely straightforward. When we compare a GAM to a parametric logit through a likelihood ratio test, we are making a comparison of the in-sample model fit. As we have demonstrated repeatedly, a GAM will often provide a better in-sample fit than a comparable GLM. For a matching analysis, however, the in-sample fit of the model is irrelevant if balance is not achieved. With matching, balance is the performance metric that matters, not in-sample fit. As such it is possible that while a GAM may provide better in-sample model fit, a parametric model may achieve better balance. Should this occur, one would prefer the parametric propensity score model to the GAM. This insight does not preclude one from using a GAM to estimate propensity scores, but one cannot assume the better in-sample fit from a GAM will necessarily translate into better balance.

Education and Wages

The data for the first example are from Dehejia and Wahba (1999) and are part of a randomized job training experiment called the National Supported Work Demonstration Program (NSW) in the mid-1970s. These data have been used by a many of researchers to test how well various nonexperimental estimators recover the experimental results.

The data set consists of 445 males and contains measures of age, education in years, indicators for whether the respondent was African-American, Hispanic, married, without a high school degree, measures of earnings in 1974 and 1975 before the job training, indicators for whether earnings were zero in 1974 and 1975, and an indictor for whether the respondent received job training, the treatment. To estimate the propensity score, we used a logistic regression model to regress the treatment indicator on all the variables as well as quadratic transformations for age, education, and wages in 1974 and 1975. We estimated a second propensity score model using a GAM with smoothing splines and GCV selected smoothing for age, education, and the wage variables. We also compared the two models with a likelihood ratio test and the GAM provides a modestly better fit ($p = 0.10$). After one-to-one matching with replacement, we evaluated the matched treatment and control groups for balance using a variety of measures including t-tests and bootstrapped Kolmogorov–Smirnov (KS) tests. We also include interactions between several of the variables in the balance tests. For the matching and balance tests, we used the algorithms in the R package `Matching` (Sekhon 2007).

Table 7.1   GLM and GAM balance test comparison wage data.

	GLM		GAM	
	Unmatched	Matched	Unmatched	Matched
Age	0.568	0.582	0.531	0.952
Education	0.016	0.335	0.019	0.684
Black	0.647	0.577	0.647	0.134
Hispanic	0.064	0.414	0.064	0.285
Married	0.334	0.467	0.334	0.824
H.S. degree	0.002	0.491	0.002	0.047
Earnings, 1974	0.585	0.252	0.568	0.484
Earnings, 1975	0.066	0.164	0.047	0.414
Unemployed, 1974	0.330	0.522	0.330	0.608
Unemployed, 1975	0.068	0.465	0.068	0.270
Earn '74 × Earn '75	0.337	0.389	0.278	0.537
Age × H.S. degree	0.040	0.814	0.026	0.429
Education × Earnings '74	0.603	0.200	0.595	0.573
Education × Earnings '75	0.076	0.165	0.048	0.856

Note: Cell entries are either t-test or bootstrapped KS test p-values for balance test

In Table 7.1, we report the results from the balance tests for both models. We only report one set of measures of balance though, in general, one should use multiple measures to assess balance. Close observation reveals that the propensity scores from the GLM are better balanced than those from a GAM. It is true that for some of the smoothed variables in the GAM, the balance is better. For example, under the GLM the balance on years of education improves from 0.016 to 0.335, while for the GAM the balance improves from 0.019 to 0.684. The GAM model, however, fails to balance the indicator for a high school degree. The initial p-value before matching is 0.002 and it is 0.047 after matching. The balance for the indicator of race also worsens under the GAM as the p-value goes from 0.647 to 0.134.

In this illustration, the logit model provided balance that was superior to the GAM, despite the fact that the likelihood ratio test indicated that the GAM provided better fit to the data. While most of the variables balanced under both propensity score models, for the GAM a key covariate failed to achieve balance. For the logit model, all the variables were balanced. In this example, then, the propensity scores from the logit model allowed for better balance than those from the GAM. One might assume from this example that one should simply use a parametric model for the estimation of propensity scores. The next example demonstrates that such a conclusion would be premature.

Presidential Campaign Visits

Keele, Fogarty, and Stimson (2004) attempt to test whether presidential campaign visits in 2002 were effective in boosting support for Republican candidates. They use propensity scores to match Congressional districts that George W. Bush visited to Congressional districts that were nearly identical along a set of observable characteristics. We replicate their analysis and generate propensity scores using a set of Congressional district characteristics to predict the propensity to treat. To generate propensity scores, we regressed the treatment indicator of whether the president visited that Congressional district or not on variables that measure the percentage of the district that is African-American, Hispanic, and urban, the percentage of those in the district that have a high school degree, the percentage with a college degree, median income in the district, a measure of the amount spent in the race by both candidates, a dummy variable for whether it is an open race, and a dummy variable for whether *Congressional Quarterly* designated the race as competitive, and the Republican presidential vote share for the district in 2000. We generated propensity scores using both a logit model with several transformations and a GAM. After one-to-one matching with replacement, we checked for balance using the same set of tests as in the last example. The results from the balance tests are in Table 7.2.

The results for this set of balance tests are quite different from the last set. For this example, the balance is much better with the GAM than with the GLM. For example, the urban measure fails to balance under the GLM with a p-value of 0.002 for the unmatched sample and a p-value of 0.043 for the matched data. Under the GAM, the p-value for the matched data is 0.782. For almost all the measures, the GAM provides higher p-values for the balance test. For example, the measure of high school education p-value for the matched data is 0.096 and 0.961 for the GLM and GAM respectively.

One might assume that the flexible specifications that are possible with GAMs would make the model a natural choice for the estimation of propensity scores. However, as the examples in this section demonstrate that is not always the case. The results from the two examples here indicate that researchers should consider GAMs as a possibility for propensity scores, but a GAM may not improve balance over GLM estimates. When one should opt for a parametric versus a semiparametric specification is an open question as research into the best practice of propensity scores and matching is currently ongoing. Analysts should perhaps

Table 7.2   GLM and GAM balance test comparison presidential visit data

	GLM		GAM	
	Unmatched	Matched	Unmatched	Matched
Open Race	0.005	0.247	0.005	0.254
C.Q. Hot Race	<0.001	1	<0.001	1
Spending	<0.001	0.351	<0.001	0.782
% Urban	0.002	0.043	0.002	0.788
% Black	0.777	0.183	0.781	0.333
% Hispanic	<0.001	0.801	<0.001	0.541
% H.S Degree	0.334	0.096	0.323	0.961
% College Degree	0.022	0.212	0.003	0.979
Median Income	0.002	0.104	0.001	0.484
2000 Pres. Vote	0.041	0.151	0.042	0.275

Note: Cell entries are either t-test or bootstrapped KS test p-values for balance test

start with a parametric model and if the balance results are not adequate alter the propensity score model specification using smoothing terms.

7.4   Conclusion

In this chapter, we outlined three more specialized applications for nonparametric regression models. First, we demonstrated that the semiparametric regression paradigm can be extended to mixed models. Besides providing for nonlinearity with mixed models, smoothers allow for flexible estimates of the interactions common to these models. Analysts can also use Bayesian techniques to estimate semiparametric regression models. The Bayesian framework allows for transparent programming of the smoothing and the estimation of confidence bands that fully reflect our uncertainty about smoothing. Finally, GAMs can also be used to estimate propensity scores for matching analyses of observational data. While GAMs can improve balance, the better in-sample fit they provide does not guarantee improved balance.

8

Bootstrapping

While nonparametric and semiparametric regression models relax functional form assumptions, they still rely on standard inferential tools such as confidence intervals and p-values. In many instances, however, the test statistics for these models do not have exact referent distributions. Smoothing splines and automatic smoothing techniques are valuable since they reduce guesswork about how much smoothing to apply and the chance of overfitting. Both of these techniques, however, complicate attempts to make inferences, and estimated p-values must be treated with care if exact inferences are important. Under these circumstances, simulation methods provide more accurate inferences. Bootstrapping is a nonparametric resampling technique used to estimate statistical uncertainty in contexts where parametric assumptions cannot be met. Bootstrapping, which has many applications beyond the scope of this book, provides a simulation method for estimating p-values that allows for testing between parametric and semiparametric fits without having to assume the test statistic follows a specific parametric distribution. With the bootstrap test, it is possible to estimate a p-value for the test between two models regardless of what method has been used for the smoothing. Hence, the chapter first discusses the bootstrap as a general inferential technique and then applies the bootstrap to nonparametric and semiparametric regression models. Our overview of the bootstrap is necessarily brief, and readers should consult Efron and Tibshirani (1993) and Davison and Hinkley (1997) for more in-depth coverage.

8.1   Classical Inference

We start with a very brief review of basic methods of statistical inference. One goal of statistical inference is to make statements about populations using random samples. Statistical inference requires measures of statistical uncertainty and a set of distributional assumptions for the calculation of confidence intervals, t-statistics and p-values. For example, recall the process for testing a hypothesis about a sample mean. Assume the analyst wishes to test whether his or her estimate is different from 0. One method for testing this hypothesis test is to use interval estimation. To conduct such a test under the principles of classical hypothesis testing, we form a confidence interval around the estimate and observe whether zero lies inside the interval. How do we form this confidence interval? It is the sample average $\pm$ 2 times the standard error.

The classical confidence interval requires the assumption that the estimate of the sample mean follows a normal distribution. Using the normal distribution critical value in the estimation of the confidence interval is an explicit acknowledgement of the parametric assumption required for classical statistical inference. Often such parametric assumptions are justified and verifiable, particularly when the sample size is large. There are a number of situations, however, where the distributional properties of the estimator are in question or are unknown. A ratio of means is an example of one statistic where the distributional properties are uncertain. When the distributional properties of the estimator are uncertain, bootstrapping can be used to derive confidence intervals. Bootstrapping is a nonparametric approach to statistical inference that substitutes simulation for the traditional distributional assumptions found in classical methods. Bootstrapping is nonparametric in that it does not assume that the underlying distribution of an estimator belongs to one of the parametric family of distributions such as the Normal, t, or F among others.

8.2   Bootstrapping – An Overview

Assume we have produced a statistical estimate, but perhaps the sample size is small or we have used a statistic for which there is no known sampling distribution. Either situation is problematic for classical inferential theory.

One solution is to bootstrap the data by taking several *new* samples from the original sample and estimating the statistic for each new resample. If we repeat this process enough times, we can form an empirical sampling distribution for the estimate, and we use this empirical sampling distribution to construct confidence intervals for the estimate. We apply the principles of inference but treat our sample as the population. With bootstrapping, the population is to the sample, as the sample is to the bootstrap samples. There are several types of bootstrapping:

- The nonparametric bootstrap: No underlying population distribution is assumed. This is the most commonly used method.

- Smoothed bootstrap: The sample distribution is smoothed before we sample from it.

- Parametric bootstrap: This method is equivalent to maximum likelihood. It assumes that the statistic has some underlying parametric form (Normal, Bernoulli, exponential, etc.). While it may be useful in some contexts, generally the imposition of a parametric assumption defeats the purpose of the bootstrap (Efron and Tibshirani (1993).

- Bayesian bootstrap: The resampling scheme instead of being random is chosen according to some prior.

Another technique that is closely related to the bootstrap is the jackknife. The jackknife also relies on resampling from the data, but each resample is composed of the original data with one observation randomly left out. Other than the different resampling scheme, one proceeds exactly as with the bootstrap. Simulation studies have shown, however, that while the jackknife performs well for computing means, the bootstrap is generally superior for computing standard errors (Efron and Tibshirani 1993). We introduce the basic principles of bootstrapping and then introduce some concepts from bootstrapping regression models.

8.2.1   Bootstrapping

To perform a bootstrap, assume we have a set a sample of data, X, with n observations. We, first, define a random variable X^*, which is composed of a series of resamples from the original data set. We treat the sample as our population and sample from it B times. Each new bootstrap sample is a random selection from the original sample. The figure below compares the traditional sampling method with the bootstrap resampling process when n is five.

Original process			
Population	$\rightarrow$	$(x_1, x_2, x_3, x_4, x_5)$	$\rightarrow$ $\bar{x}$
Bootstrap resampling process			
BS Sample 1	$\rightarrow$	$(x_3, x_2, x_3, x_4, x_5)$	$\rightarrow$ $\bar{x}_1^*$
BS Sample 2	$\rightarrow$	$(x_1, x_2, x_2, x_4, x_5)$	$\rightarrow$ $\bar{x}_2^*$
$\vdots$			
BS Sample B	$\rightarrow$	$(x_1, x_4, x_2, x_4, x_5)$	$\rightarrow$ $\bar{x}_B^*$

The sampling is done with replacement, meaning each new bootstrap sample may contain more than one realization of the data points from the original sample. Using basic combinatorics, the possible number of resamples is n^n. Thus, it is easy to see that the number of possible resamples grows quickly with the sample size. For example, with a sample size of four, there are only $4^4 = 256$ possible resamples each with a probability of 1/256. Due to sampling with replacement, however, most of the samples will contain repeated observations in them. With a sample size of four with 256 resamples, only 100 of the samples will not have repeated observations. Note that increasing the sample size by one to five increases the number of possible resamples to 3125, making it impractical to enumerate all the possible resamples.

After each resample, we evaluate the statistic of interest. In this example, we assume that statistic is the sample mean, so we calculate the mean for each resample. The set of means from each resample forms an empirical sampling distribution for the statistic. We use this sampling distribution to calculate standard errors and confidence intervals that do not rely on a parametric distribution. The bootstrap standard error is

$$\widehat{SE}^* = \frac{\sqrt{\sum_{b=1}^{B}(\bar{x}_b^* - \hat{\theta}^*(\cdot))^2}}{B-1} \tag{8.1}$$

where $\hat{\theta}^*(\cdot)$ is

$$\sum_{b=1}^{B} \bar{x}_b^*/B, \tag{8.2}$$

the mean of all the bootstrap resamples. One should correct for small samples by multiplying the bootstrap standard error by $\sqrt{\frac{n}{n-1}}$ (Efron and Tibshirani 1993).

In sum, we apply the bootstrap by sampling with replacement from the data. For each resampled data set, we recalculate the mean and save it. After we have done this many times, we form a new empirical sampling distribution composed of the repeated mean estimates. The bootstrapped standard error is the square root of the sum of squares of each resample mean, less the mean of all the resamples, divided by the number of bootstrap samples. In short, the distribution of bootstrap means becomes the previously unknown sampling distribution of the original sample. How many resamples do we need? The standard rule of thumb is that 200 resamples are needed for a standard error, and 1000 resamples are needed for a confidence interval (Efron and Tibshirani 1993), but there is no one right answer. One method is to run a reasonable number of resamples, and then change the random number seed and repeat the process. If the results change in a meaningful fashion increase the number of resamples. Once the bootstrap resampling procedure is completed, one can then estimate confidence intervals. There are three different bootstrapping confidence intervals.

Bootstrap Confidence Intervals

Normal-Theory Intervals The simplest bootstrap confidence interval is the Normal-Theory interval, which uses the the bootstrap standard error with the classical method of calculating confidence intervals. The Normal-Theory confidence interval formula is:

$$CI_{(\theta)} = \hat{\theta} \pm z_{\frac{\alpha}{2}} \widehat{SE}^*. \tag{8.3}$$

The term $z_{\frac{\alpha}{2}}$ refers to the z-score from the normal distribution, and $\widehat{SE}^*$ is the estimated bootstrap standard error. The Normal-Theory approach works well when the bootstrap sampling distribution is approximately normal. One can test this assumption with a quantitle-quantitle (Q-Q) plot. The bootstrap, however, is designed to be less reliant on distributional assumptions, and it makes little sense to return to the parametric approach at this point.

Bootstrap Percentile Intervals

The percentile confidence interval uses the percentiles of the bootstrap sampling distribution as an estimate of the confidence interval. This approach requires larger bootstrap samples than the Normal-Theory method. The analyst should use at least 1000 resamples for calculating these confidence intervals as opposed to the 200 used to compute the standard error. For large bootstrap samples, the confidence interval is bounded at the $\alpha/2$ and $1 - \alpha/2$ percentiles:

$$\hat{\theta}^*_{lower} < \theta < \hat{\theta}^*_{upper} \tag{8.4}$$

where $\hat{\theta}^*_{lower}$ and $\hat{\theta}^*_{upper}$ are the ordered bootstrap replicates. For a standard 95% confidence interval, the value of $\hat{\theta}^*_{lower}$ would be 0.025 and the value of $\hat{\theta}^*_{upper}$ would be 0.975. Calculating percentile intervals is a matter of reordering the bootstrap distribution from the smallest to largest values and calculating the percentile values in the distribution. For example, assume we have a bootstrap sampling distribution of 1000 values. For a 95% confidence interval, we use the $1000 \times \alpha/2 = 25$th values from the ordered bootstrap values as the confidence interval endpoints.

Bias-corrected, accelerated confidence interval, BC_a

The bias-corrected, accelerated, or BC_a, confidence interval adjusts the confidence intervals for bias by employing a normalization transformation through two correction factors: Z and A (Efron and Tibshirani 1993). These CIs correct for small sample bias and are acceleration adjusted (acceleration is the rate at which the standard errors converge as N increases). Both of these corrections fix flaws in the Normal-Theory and percentile confidence intervals. While both of these confidence intervals often work, they have less than desirable coverage properties (Efron and Tibshirani 1993). The corrections in the BC_a provide better coverage than the other two methods for confidence intervals under a wider set of situations. The correction factors Z and A are defined below. The first, Z, is:

$$Z = \Phi^{-1}\left[\frac{\#(\hat{\theta}^*(b) < \hat{\theta})}{B}\right] \tag{8.5}$$

where Φ^{-1} is the inverse of the standard normal CDF. In short, Equation (8.5) is the number of bootstrap replicate estimates that are below the original sample parameter estimate (the mean in our running example) divided by the number of bootstrap samples. Roughly, Z measures the discrepancy between the median of $\hat{\theta}^*$, the bootstrap sampling distribution, and the median of the original sample. The estimate of Z will be 0 if exactly half of the bootstrap sampling distribution values are less than or equal to the parameter estimated in the original sample.

The formula for A is more complex. Let $\mathbf{X}_i$ be the original sample with the ith case deleted and let $\hat{\theta}_i$ equal the statistic of interest applied to this new sample. Finally, define $\hat{\theta}_{(\cdot)} = \sum_{i=1}^n \hat{\theta}_i / n$. With these definitions, we define A as:

$$A = \frac{\sum_{i=1}^n (\hat{\theta}_{(\cdot)} - \hat{\theta}_i)^3}{6\left[\sum_{i=1}^n (\hat{\theta}_{(\cdot)} - \hat{\theta}_i)^2\right]^{\frac{3}{2}}}. \tag{8.6}$$

The correction factor, A, causes the standard error to grow as the sample size decreases. We combine A and Z to calculate the lower and upper BC_a CIs.

$$A_{\text{lower}} = \Phi\left[Z + \frac{Z + z_{(\alpha/2)}}{1 - A(Z + z_{(\alpha/2)})}\right] \tag{8.7}$$

$$A_{\text{upper}} = \Phi\left[Z + \frac{Z + z_{(1-\alpha/2)}}{1 - A(Z + z_{(1-\alpha/2)})}\right]. \tag{8.8}$$

Φ is the cumulative standard normal distribution function.

When discussing the performance of confidence interval estimation, CIs can be first order accurate and second order accurate. First order accuracy implies that the confidence interval is transformation respecting, which implies that the confidence interval is invariant to any mathematical transformation performed on the parameter of interest. Second order accuracy implies that errors in the confidence interval go to zero at a rate of $1/n$. Confidence intervals that are both first and second order accurate have the best coverage properties. The BC_a CIs are both first and second order accurate, while the Normal-Theory and percentile confidence intervals are not. See Efron and Tibshirani (1993) for a full discussion of bootstrap confidence intervals and their properties.

8.2.2 An Example: Bootstrapping the Mean

Bootstrapping is best understood through an example. Table 8.1 contains hypothetical data on the seat shares for two parties in a legislature across five elections. Party 2 has maintained a clear majority across four of the elections. The analyst may wish to estimate the average difference in seats held, and a confidence interval around that point estimate. With such a small sample, it is hard to justify the distributional assumption needed for a classical confidence interval, and, therefore, the standard error may not be accurate. Under these circumstances, the bootstrap provides an alternative method for estimating the confidence interval.

Using classical methods, we can easily calculate the mean, the standard deviation, and confidence intervals. The estimated mean for the sample is 24.75, and the estimated standard error is 11.15. These are calculated in the standard way: $\bar{Y} = \sum Y_i/n = (13 + 31 + 37 + 18 + -3)/5$, and $S = \sqrt{\sum(Y_i - \bar{Y})^2/(n-1)}$. We next calculate a 95% confidence interval; given the small sample size, we use a t-distribution:

$$CI = \bar{Y} \pm t_{n-1, 0.025} \times \hat{\sigma}\sqrt{n}$$
$$= 19.20 \pm 2.78 \times 15.72$$
$$= 19.20 \pm 43.70. \tag{8.9}$$

Notice that the confidence interval is rather wide and includes zero, which implies that we would be unable to reject the null hypothesis that the sample mean is different from zero.

Table 8.1 Hypothetical data on party seat shares.

	Party 1	Party 2	Difference
Election 1	208	221	13
Election 2	200	231	31
Election 3	199	236	37
Election 4	208	226	18
Election 5	219	216	−3

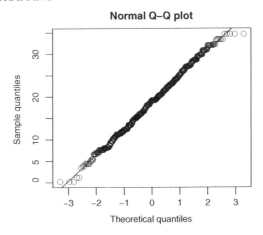

Figure 8.1   Q-Q plot of the bootstrap sampling distribution.

Now, for comparison, we bootstrap the data. We estimate the bootstrap sampling distribution by resampling 1000 times from the five seat differences. For each resample, we calculate the mean, and the bootstrap sampling distribution is composed of this set of means. The mean of the bootstrap sampling distribution is the bootstrap estimate of the mean. In this example, the bootstrap estimate of the mean is 18.68, which is similar to the classical estimate. Using Equation (8.1), we calculate the bootstrap estimate of the standard error. The bootstrap standard error estimate in this case is 6.29, and the sample size corrected bootstrap standard error is 7.03. Both of these estimates of the standard error are much smaller than the classical estimate of 15.72, which indicate that our estimate is more precise than if we assume the data a drawn from a *t*-distribution, making it less likely that we will be unable to reject the null hypothesis.

Next, we use the bootstrap estimate of the standard error to calculate Normal-Theory confidence intervals. The Normal-Theory interval assumes that the bootstrap resamples are normally distributed. We can evaluate this assumption using a Q-Q plot. Figure 8.1 contains a Q-Q plot of the bootstrap sampling distribution, and there are minor departures from normality in one of the tails. Therefore, we should calculate both percentile and bias corrected confidence intervals to ensure that the results do not depend on a normality assumption.

For comparison, we estimate the Normal-Theory confidence interval which is 4.91 and 32.46.[1] The Normal-Theory confidence interval is shorter than the classical estimate since the bootstrap standard error is smaller. With the Normal-Theory

---

[1] We used the sample size adjusted standard error to calculate the Normal-Theory confidence interval.

CI, we are able to reject the null hypothesis. Estimating percentile confidence intervals requires ordering the bootstrap sampling distribution and identifying the appropriate cases that correspond to the 2.5 and 97.5 percentiles. For this example, the percentile confidence interval is 7.0 and 31. While slightly different, the percentile confidence interval does little to change our general inference. The BC_a confidence interval is 5.0 and 29.8. This example is extremely simple, but it provides an introduction to how bootstrapping can be used to provide statistical inferences without parametric assumptions. Now, we introduce some concepts from the bootstrapping of regression models.

8.2.3   Bootstrapping Regression Models

The bootstrap can also be applied to regression models. Inferences for regression models depend on several assumptions such as homoskedasticity and normality of the error term. Bootstrapping is useful for least squares regression models when we suspect that such assumptions have been violated. Bootstrapping is also suited to other models where it is difficult to derive analytic formulas for the standard errors. This is true for some robust regression models and quantities such as predicted probabilities from a GLM.

There are two different bootstrapping algorithms for regression models. The first is called random pairs or random-X bootstrapping. Here, we resample from the data and estimate the regression model using the bootstrap samples. In this approach, the regressors are treated as random. The other method is called residual bootstrapping or fixed-X bootstrapping. This approach involves forming a new dependent variable from the model residuals and then regressing the constructed Y on X to obtain the bootstrap replications of the coefficients. In this approach, the regressors are treated as fixed, which implies that the regression model fit to the data is correct. An outline of the algorithm for each method follows.

Bootstrapping Pairs   The bootstrapping pairs algorithm is a generalization of the method we used to bootstrap the mean.

1. Form a bootstrap resample by sampling paired Y's and X's. The size of the resample should be equal to the original sample size N, but sampling is done with replacement.

2. Estimate the regression on the bootstrap resample.

3. Save the estimates from this regression.

4. Repeat Steps 1 through 3 B times taking a new bootstrap sample each time, where B is the total number of bootstrap replications.

We now have a bootstrap sampling distribution for each of the $\hat{\beta}$ coefficients. With this sampling distribution, we can calculate standard errors and confidence intervals in exactly the same way as we did for the mean.

Bootstrapping Residuals   Bootstrapping residuals differ in several respects from resampling pairs.

1. Estimate the regression model and form the residuals $(\hat{\varepsilon} = \hat{Y} - \mathbf{X}\hat{\beta})$.

2. Take a bootstrap sample of the residuals (a sample of size N error terms with replacement:

$$\varepsilon^* = (\varepsilon_1, \varepsilon_1, \varepsilon_3, \ldots, \varepsilon_n). \tag{8.10}$$

3. Generate a set of bootstrap Y^*'s in the following way:

$$Y^* = \hat{Y} + \varepsilon^*. \tag{8.11}$$

4. Y^* is then regressed on $\mathbf{X}$ to form a vector of $\hat{\beta}^*$'s

5. Steps 2–4 are repeated B times, where B is again the number of bootstrap replications.

Each vector of the $\hat{\beta}_k^*$s is the bootstrap sampling distribution for that parameter. Again, standard errors and confidence intervals are estimated using this sampling distribution.

Residuals versus Pairs   The reader might wonder whether one should prefer resampling of pairs or residuals. There are advantages to both, but the answer depends on how much faith one places in the specification of the regression model and the context in question. When we resample residuals, it implies that the specification of the model is correct and that the error term is not dependent on the form of $\mathbf{X}$ (Efron and Tibshirani 1993). Depending on the context, this can be a strong assumption. Bootstrapping pairs is less sensitive to model assumptions than bootstrapping residuals, and the standard error estimates should be accurate even if the model specification is incorrect. Bootstrapping pairs, however, ignores the error structure of the regression model. One goal of resampling is to mimic the random component of the process, and the bootstrap sampling distribution better reflects the error structure when residuals are resampled. Sampling pairs also introduce additional variability into the estimates of the standard errors. There are certain contexts when the resampling pairs method is preferred for practical reasons. For models where the residuals are not easily defined (some event history models, models for binary data) bootstrapping of residuals is not always possible. Bootstrapping pairs, on the other hand, is easily applied to any model. In general, one ought try both methods and compare the results as a robustness check.

8.2.4   An Example: Presidential Elections

We provide a brief example of bootstrapping regression models. Bartels and Zaller (2001) review several statistical models of US presidential elections in the post World War II era. For one of their analyses, they regress the incumbents' vote share on real disposable income, a measure of policy moderation and the change in GDP for 14 presidential elections. They find evidence that real disposable income and policy moderation are related to incumbent vote share, which suggests that, in the aggregate, the outcomes of presidential elections are related to personal economic circumstances and candidate issue positions. With 14 cases, we ought to be hesitant to rely on standard inferential methods.

A Q-Q plot reveals that the residuals have an asymmetric distribution which implies a lack of normality. Instead of relying on parametric assumptions, we bootstrap the model and calculate confidence intervals for each estimated β parameter. The first step is to perform the resampling. For this example, we bootstrapped the model using both pairs and residuals with 1500 resamples, which implies that we will have 1500 estimates for each parameter in the regression model. Next, one should examine the bootstrap resampling distributions. Here, we use Q-Q plots to assess whether the bootstrap distributions for each β are normally distributed. Figure 8.2 contains both a histogram and Q-Q plot of the

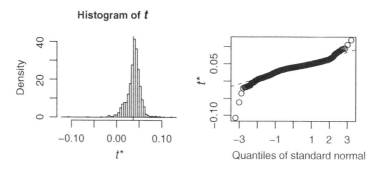

Figure 8.2   Histogram and Q-Q plot of bootstrap sampling distribution for policy moderation.

Table 8.2   Incumbent vote share: Standard and bootstrapped models.

	Classical	Residuals	Pairs
Real disposable	0.03	0.03	0.03
Income	[0.006, 0.044]	[0.011, 0.041]	[0.008, 0.047]
Policy moderation	0.04	0.04	0.04
	[0.008, 0.066]	[0.014, 0.058]	[−0.009, 0.057]
GDP	−0.001	−0.001	−0.001
	[−0.019, 0.017]	[−0.0137, 0.145]	[−0.019, 0.023]
Constant	0.44	0.44	0.44
	[0.402, 0.489]	[0.415, 0.477]	[0.404, 0.484]
N	14	14	14

Note: Cell entries are coefficient estimates.
95% Confidence intervals in parentheses.
Bootstrap confidence intervals are BC_a.

bootstrap distribution for the policy moderation variable. There are obvious departures from normality as the data in both tails show obvious departures from the benchmark of the Q-Q plot, which implies that we should use either percentile or BC_a confidence intervals.

The results with classic confidence intervals as well as the bootstrapped coefficients and confidence intervals are in Table 8.2, where we report BC_a confidence intervals. The model in the first column of Table 8.2 contains a estimates with classical confidence intervals. We see that the estimates for both real disposable income and policy moderation are both bounded away from zero, while the effect of GDP is indistinguishable from zero. The results in the second column are from the residual based bootstrap. The results are quite similar to the classical estimates, though the bootstrapped confidence intervals are slightly wider. The third column contains results based on resampling pairs. Here, we find that the confidence interval for the policy moderation variable now contains zero since the bias corrected confidence interval for this coefficient is −0.019 and 0.057. Even if bootstrapped inferences are no different, bootstrapping still serves as a check on the model assumptions in contexts where those assumptions may be violated. In this example, the results are not all identical. When the bootstrapping is based on pairs, we get results that differ from those based on classical methods or bootstrapping based on residuals.

How might we evaluate which set of confidence intervals is correct? A plot of the residuals reveals some level of misspsecification, which indicates that we ought to rely on resampling of pairs which is less sensitive to model specification.

8.3   Bootstrapping Nonparametric and Semiparametric Regression Models

In this section, we outline two different bootstrap applications for nonparametic and semiparametric models. The first application is bootstrapping standard nonparametric regression estimates.

8.3.1   Bootstrapping Nonparametric Fits

The reader may recall that the standard procedure for estimating confidence intervals for a nonparametric fit is the plug-in method. With the plug-in method, pointwise standard errors for the nonparametric fit are plugged into the standard confidence interval formula and then plotted along with the nonparametric fit. The plug-in method, of course, assumes a normal distribution of the data, which may or may not be reasonable. Moreover, such confidence intervals do not take into account bias in the nonparametric estimate. With smoothing splines, we relied on various approximations for the confidence intervals. It was only with MCMC techniques that we were able to estimate confidence intervals that fully accounted for estimation uncertainty. Bootstrapping offers another method for estimating confidence intervals without having to make assumptions about the sampling distribution of the nonparametric estimate.

To bootstrap a nonparametric regression model, we form a bootstrap sampling distribution for *each* local $\hat{y}$ value by resampling the data and estimating the nonparametric regression model for each resample. This provides a sampling distribution comprised of multiple nonparametric estimates. We calculate bootstrap confidence intervals at each of the $\hat{y}$ values that make up the nonparametric estimate. One can sample either pairs or residuals. Note that bootstrapping a nonparametric regression model is more time consuming than using the plug-in method since it requires estimating the nonparametric model multiple times.

As an illustration we estimate bootstrapped confidence intervals using the Jacobson data on challenger vote share from the 1992 House elections (Jacobson and Dimock 1994). Recall that using these data in Chapters 2 and 3, we found that support for Perot in a district had a nonlinear effect on challenger's vote share. We again estimate a nonparametric regression model for challenger's vote share and the percentage of the vote that Perot received in the 1992 presidential election using smoothing splines with the amount of smoothing estimated via GCV. We estimated confidence bands with the plug-in method and with the bootstrap. Figure 8.3 contains the nonparametric fit between the challenger's vote share and support for Perot along with both standard and bootstrapped percentile confidence

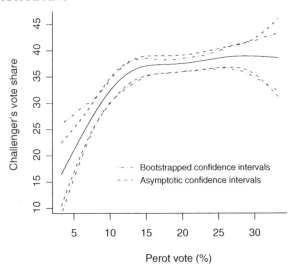

Figure 8.3   Plug-in and bootstrap confidence intervals for effect of Perot vote on challenger vote share.

bands. The differences between the two confidence bands are negligible, and our inferences about the nonlinear effect remain the same.

While bootstrapping the confidence bands makes little difference in this particular instance, bootstrapping allows the analyst to estimate confidence bands in situations where one suspects that the distributional assumptions do not hold. Moreover, bootstrapping also provides a means of estimating confidence bands that take smoothing uncertainty into account without using MCMC simulation. Bootstrapping can be applied in a similar way to additive and generalized additive models if one wants estimates of the confidence bands that do not rely on distributional assumptions.

8.3.2   Bootstrapping Nonlinearity Tests

One component of inference for both nonparametric and semiparametric regression models is testing the spline or local polynomial estimate against a parametric alternative. For smoothing splines the distribution for the test statistic is only approximate, and when automated smoothing methods are used less is known about the distribution of the test statistic. The use of overparameterized cubic splines is one method that ensures the p-value estimates are accurate. Bootstrapping also allows for testing a parametric null model against a semiparametric model. By bootstrapping the test statistic of the two models being compared, one can calculate a p-value that does not rely on a distributional assumption. This is useful when hypothesis tests are of primary concern and the estimated p-value is near the threshold of interest, particularly when one has used either smoothing splines or automated smoothing.

Hastie and Tibshirani (1990) first described how bootstrapping may be used to test between parametric and semiparametric models, but Davison and Hinkley (1997) provide an explicit algorithm for the test. Importantly, the algorithm requires resampling residuals and not cases. Since we are testing against the parametric model, it needs to be the basis of the resampling (Davison and Hinkley 1997). By resampling residuals, we capture the error structure of the null model that we are testing the semiparametric model against. The steps in the algorithm are:

1. Estimate the null model (a parametric model).

2. Estimate the semiparametric model.

3. Calculate t, the test-statistic from a likelihood ratio test between the models.

4. Estimate the model residuals, $\hat{\varepsilon}$ from the null model.

5. Resample with replacement from the residuals to form ε^*.

6. Form $Y^* = \hat{Y} + \varepsilon^*$.

7. Estimate both the parametric and semiparametric models using Y^* as the dependent variable.

8. Calculate t^*, the test statistic from the models fit to Y^*.

9. Repeat B times.

10. Calculate the new p-value using the following formula:
$$\frac{[1 + \#(t^* \geq t)]}{(B + 1)}$$

where $\#(t^* \geq t)$ is the number of times t^* is equal to or exceeds the original t-statistic t.

The algorithm returns an easily interpretable p-value that can be compared to the p-value from the likelihood ratio test. The bootstrapped test tends to be more conservative, as we might expect. If the hypothesis test for nonlinearity is critical, it is a good practice to check the standard test against a bootstrapped test. For general diagnosis, however, the standard likelihood ratio test remains a useful guide.

The test results reported in this text thus far have relied upon standard likelihood ratio tests. As an illustration, we revisit one of those tests and bootstrap it. In Chapter 5, using Jacobson's data on the 1992 House elections, we estimated a parametric model for challenger's vote share as a function of whether or not the challenger is experienced, challenger and incumbent spending, presidential vote in the district, support for Perot, whether or not the district was marginal, whether or not the district had been redistricted, and the number of overdrafts on the House Bank. In the example in Chapter 5, all the continuous variables were tested for nonlinearity, but here, we focus on the test for the challenger spending variable only.

The process starts with estimating the semiparametric model of interest. We estimated a semiparametric model using smoothing splines and GCV to estimate the amount of smoothing to be applied rendering the distribution for this test statistic approximate. Under a standard χ^2 test, the p-value for the test statistic is highly significant ($p < 0.001$). As a comparison, we estimated a third model using natural cubic splines with six knots and repeated the likelihood ratio test. The model with splines remains a better fit to the data ($p < 0.001$). Finally, we estimated the p-value using the bootstrap. We used 999 bootstrap resamples following the above algorithm. Again, we find that the spline model is a better fit to the data than the parametric model ($p = 0.001$). The bootstrap algorithm provides a p-value estimate that does not rely on a distributional assumption. The bootstrap p-value is more conservative than either of the tests that rely on the χ^2 distribution, but the test still clearly indicates that the semiparametric model is a better fit.

8.4   Conclusion

Bootstrapping allows analysts to obtain estimates of statistical uncertainty with simulation, eliminating the need to invoke parametric probability distributions that may or may not be appropriate. While bootstrapping was once considered a computationally intensive process, the speed of modern computers means that the process takes much less time than it once did. For example, the time required to estimate the bootstrap p-value in the Congressional elections example above was less than five minutes on a standard laptop. While bootstrapping has a number of useful applications, in the context of semiparametric regression, it allows the analyst to test between parametric and semiparametric models without having to assume that the test statistic follows a specific distribution. This is particularly useful when automated smoothing techniques are used, when even less is known about the distribution of the test. When hypothesis testing is of interest, the bootstrap provides a useful check of the parametric assumptions.

8.5 Exercises

Data sets for exercises may be found at the following website: http://www.wiley. com/go/keele_semiparametric.

1. Use the data in `allmen1986.dta` from the exercises in Chapter 5. Replicate the analysis from the second example in Chapter 5 using this data set. Estimate the model with penalized splines and automated smoothing. Use bootstrapping to estimate the p-values for the tests. Compare your results to likelihood ratio tests from models with smoothing splines and natural cubic splines.

2. Use the `forest.dta` data on evironmental degradation of forests across countries. Model deforestation as a function of the following variables: `dem`, `wardum`, `rgdpl`, `openc`, and `popdense`.

   (a) Fit a semiparametric regression model to the data using spline fits for all the continuous predictors. Plot all the nonparametric estimates. Which relationships appear to be nonlinear?

   (b) Use bootstrapping to test which terms are nonlinear.

   (c) Fit a nonparametric regression model between the outcome and one of the variables that has a nonlinear effect and bootstrap the confidence intervals.

3. The data set `latamerturnout.dta` contains data on voter turnout from all minimally democratic elections from 1970 to 1995 in Latin America. Develop a model of turnout. Decide which terms should be modeled parametrically and which should be modeled nonparametrically using bootstrapped nonlinearity tests.

9

Epilogue

We end with three caveats. First, semiparametric models are not a panacea. The ability to flexibly model nonlinearity within the context of standard statistical models is a useful tool, but it is not a substitute for careful analysis of data. A well thought-out specification that is motivated by strong theory within a fully parametric model is preferable to a semiparametric regression model with a poor specification. The reader should not think that semiparametric models are always 'better' than standard parametric models. The semiparametric model should be thought of as another model in the analyst's toolkit.

Frequently, the semiparametric model will only serve as a diagnostic tool. Nonparametric techniques provide the best means of testing for nonlinearity in standard models, and if relationships are truly linear or if data transformations are adequate, there is little need for the semiparametric model. As a diagnostic tool, however, the semiparametric model urges us to take the possibility of nonlinearity more seriously. Currently, we suspect that many analysts do not as a matter of course test for whether transformations are required on the right-hand side of their models. Nonlinearity is common enough and easy enough to test for that analysts ought to ensure that any continuous covariate is tested for nonlinearity just as one should always test for unequal error variances or outliers. Nonparametric techniques allow the analyst to flexibly test for nonlinearity and test how well power transformations model that nonlinearity.

The second caveat revolves around model complexity. At one time, nonparametric and semiparametric regression models were classfied as computationally intensive statistical models. The power of computers has increased so dramatically that one can now fit a semiparametric model in almost the same time as a linear model, despite the iterative algorithm that is required for estimation. Given the ease of estimating semiparametric models, it is easy to forget that they are asking more of the data than assuming a fully parametric specification. In some cases, the addition of a nonparametric term to a model results in convergence warnings or a complete failure to converge. This is typically not a problem for simpler models such as the linear model or for most GLMs. It does, however, become increasingly problematic for more complex models. For example, we have found the use of a semiparametric specification in conjunction with frailties in a Cox model will often fail to converge. Convergence problems are also more common when fitting semiparametric mixed models.

What is to be done in such situations? In both of the above examples, it would appear that the analyst must choose between modeling unobserved heterogeneity and nonlinearity. There are no means for deciding which of these maladies is more deserving of a cure as both are specification errors. As a practical matter, one might model the nonlinearity alone and assess how well transformations capture the nonlinearity. If the transformation improves the model fit, the transformed data could be used with the model for heterogeneity. While one may not be able to use the nonparametric fit in the final model, it can at least be used for diagnosis during an earlier stage of data analysis.

Finally, analysts should always use some caution in the interpretation of the results from nonparametric regression estimators. One should assume that high degrees of nonlinearity are the result of overfitting. While smoothing splines make such fits rare, they are not impossible. Additional smoothing will iron out such nonlinearity, and interpretation should concentrate on the general form of the nonlinearity as opposed to local variation.

Appendix: Software

The data analysis for every example in this book was conducted in R, a GNU implementation of the S language, and Stata. Stata was only used for data management. While there are other statistical packages that can be used (SAS and S-Plus come to mind) for nonparametic and semiparametic regression models, R provides the best platform for the estimation of these models. The S language has always had excellent capababilities in this area as Hastie and Tibshurani wrote the software that accompanies their seminal text in S. It is also true that while many standard statistical packages like Stata have the ability to estimate nonparametric regression models, few can estimate semiparametric regression models. Finally, our use of Stata for data management is completely a function of habit and familiarity. Many other software packages could be used for data management, since R can read a variety of data formats including SAS and SPSS.

Here, we provide a brief introduction to semiparametric regression models in R. This introduction is not meant to be a comprehensive review of R or of semiparametric models in R. The reason we do not provide any coverage on the basics of R is that there are several good books readily available on this topic. I would recommend books by Venables and Ripley (2002) and Maindonald and Braun (2007) for all around introductions to R. I do not provide a comprehensive treatment of semiparametric regression models in R because in truth R is more like series of statistical software packages for nonparametric and semiparametric regression models. For example, there are at least four different interfaces for the estimation of smoothing splines. The reason for this is that much of the ability to estimate specialized routines in R is the result of user contributed libraries called packages that add various capabilities to the basic functionality of R. At the time of writing there were at least six different R packages for the estimation of semiparametric regression models.[1] All the semiparametric models estimated in this text were estimated with either the mgcv or SemiPar pacakages, though mostly with mgcv.

Regardless of which package one uses, there are two steps in the use of the software. The first is the basic estimation of the model, and the second is interacting with the graphics functions to view the results and later to produce publication quality plots. A typical semiparametric regression model equation in R would take the following form:

```
gam(y ~ x1 + x2 + s(x3), data=data.set)
```

The ~ term can be read as = or y is a function of. The other parts of the model equation are:

- x1 + x2,... predictor variables to be modeled parametrically.

- s(x3) indicates that this predictor variable should be modeled nonparametrically. Some packages use s() to designate nonparametric estimation, while others use f(). Depending on the package, use of s() or f() without any arguments can imply either that smoothing is done according to default settings or automatic smoothing selection techniques are used. For example, in mgcv the default is to use generalized cross-validation for the smoothing parameter selection. Almost always, users can set the amount of smoothing, choose among different smoothing parameter selection methods, and change the basis functions.

- data= points the estimation routine to the data set to be used. This is important since R can hold several data sets in memory unlike many statistical software programs such as Stata.

- The above model equation would default to linear regression. To estimate a logistic regression, instead, the analyst must add family=binomial. For a Poisson regression model, the analyst would add family=poisson.

One inconvenient wrinkle is that while most generalized semiparametric packages can estimate several different types of models, none are comprehensive. For example, none of the R packages that we have mentioned so far have the ability to estimate the Cox proportional hazards model in any form. To estimate a Cox

---

[1]Those packages would be: gss, gam, locfit, SemiPar, vgam, and mgcv.

model, with or without nonparametric terms one must use the `survival` package. The basic functionality, however, is quite similar, in that predictor variables that are to be modeled nonparametrically one uses either `ns(x3)` for natural splines or `psplines(x3)` for smoothing splines. The vgam package extends the semiparametric framework to a number of models not supported in the more commonly used semiparametric packages. For example, with `vgam`, one can estimate semiparametric ordered logit or probit models.

While one can have R return statistical output directly to the computer screen, more often the model results are saved as an object in the following way:

```
model.1 <- gam(y ~ x1 + x2 + s(x3), data=data.set)
```

In the above model equation, the results are saved in the object `model.1`. This object is typically a list, and the analyst can extract various parts of the object list to be printed to the screen. For example, one could type `deviance(model.1)` at the command line to print only the estimated deviance to the screen. For semiparametric regression models, the model object can be passed to the R graphics function:

```
plot(model.1, se=TRUE, ylab="Some Text", xlab="Some
Other Text")
```

There is wide variation in the exact plotting parameters depending on the R package being used, but the functionality tends to be fairly similar. Most will allow the user to specify which nonparametric term to plot if there is more than one; otherwise all terms are plotted in order. The best way to learn the how to use the software, however, is personal experience. Below we reproduce the code for several of the models and figures in this book. Every data set used in this book and the code required to replicate each example is available on the book website. Example of Figure 2.13, *loess* fit with confidence bands:

```
nonpar.fit <- loess(chal.vote ~ perotvote, span=.5,
   degree=1, data=jacob)
perot <- seq(min(perotvote), max(perotvote),
   length=312)
fit <- predict(nonpar.fit, data.frame
   (perotvote=perot), se=TRUE)
plot(perotvote, chal.vote, type="n",
   ylab="Challengers' Vote Share (\%)",
```

```
xlab="Vote for Perot (\%)", bty="l")
points(perotvote, chal.vote, pch=".", cex=1.75)
lines(perot, fit$fit, lwd=1)
lines(perot, fit$fit + 1.96*fit$se.fit, lty=2, lwd=1)
lines(perot, fit$fit - 1.96*fit$se.fit, lty=2, lwd=1)
```

Example of Figure 3.8 natural cubic splines with confidence bands

```
library(splines)
mod.nspline <- lm(chal.vote~ns(perotvote, df=4,
  intercept=TRUE), data=jacob)
perot <- seq(min(perotvote), max(perotvote),
  length=312)
sfit <- predict(mod.nspline, inteval="confidence",
  se.fit=TRUE,
data.frame(perotvote=perot))
plot(perotvote, chal.vote, type="n",
  ylab="Challengers' Vote Share (\%)",
xlab="Vote for Perot (\%)", bty="l")
points(perotvote, chal.vote, pch=".", cex=1.75)
lines(perot, sfit$fit, lwd=1)
lines(perot, sfit$fit + 1.96*sfit$se.fit, lty=2,
  lwd=1)
lines(perot, sfit$fit - 1.96*sfit$se.fit, lty=2,
  lwd=1)
```

Example of Figure 3.9, the smoothing spline fit:

```
library(mgcv)
sm.1 <- gam(chal.vote ~ s(perotvote, bs="cr", k=4,
  fx=TRUE))
plot(sm.1, rug=FALSE, se=TRUE, ylab="Challengers'
  Vote Share (\%)",
xlab="Vote for Perot (\%)", residual=TRUE,
  shift=33.88, bty="l")
```

Figure 4.8 automated smoothing for challenger vote share and support for Perot

```
library(mgcv)
gam.1 <- gam(chal.vote ~ s(perotvote), data=cong)
```

```
plot(gam.1, rug=FALSE, se=FALSE, ylab="Challengers'
  Vote Share (\%)",
xlab="Number of Overdrafts", residual=TRUE,
  shift=33.88, bty="l",
main="Automatic Selection")
```

Abbreviated example of semiparametric regression model for Congressional elections from Chapter 5.

```
library(mgcv)
#Baseline Model
ols.4 <- gam(chal.vote ~ exp.chal + chal.spend.raw +
  inc.spend.raw + pres.vote + I(checks.raw^2) +
  marginal + partisan.redist + perotvote, data=cong)
#Semiparametric Model
gam.5 <- gam(chal.vote ~ exp.chal + s(chal.spend.raw,
  bs="cr") + s(inc.spend.raw, bs="cr") + s(pres.vote,
  bs= "cr") + logchecks1 + marginal + partisan.redist
  + s(perotvote, bs="cr"), data=cong)
#Model Comparison
anova(ols.4, gam.5, test="Chisq")
#Plot Results
par(mfrow = c(2,2))
plot(gam.5, select=1, rug=FALSE, se=TRUE,
  ylab="Challengers' Vote Share (\%)",
xlab="Challenger Spending", residual=FALSE, bty="l",
  shift=33.02)
points(chal.spend.raw, chal.vote, pch=".", cex=1.75)
plot(gam.5, select=2, rug=FALSE, se=TRUE,
  ylab="Challengers' Vote Share (\%)",
xlab="Incumbent Spending", residual=FALSE, bty="l",
  shift=33.02)
points(inc.spend.raw, chal.vote, pch=".", cex=1.75)
plot(gam.5, select=3, rug=FALSE, se=TRUE,
  ylab="Challengers' Vote Share (\%)",
xlab="Challenger's Party Pres. Vote", residual=FALSE,
  bty="l", shift=33.02)
points(pres.vote, chal.vote, pch=".", cex=1.75)
plot(gam.5, select=4, rug=FALSE, se=TRUE,
  ylab="Challengers' Vote Share (\%)",
```

```
xlab="Vote for Perot (\%)", residual=FALSE, bty="l",
  shift=33.02)
points(perotvote, chal.vote, pch=".", cex=1.75)
#Nonlinear Interaction
gam.7 <- gam(chal.vote ~ exp.chal + s(chal.spend.raw,
  checks.raw) +
s(inc.spend.raw, bs="cr") + s(pres.vote, bs= "cr")  +
  marginal + partisan.redist + s(perotvote, bs="cr"),
  data=cong)
#Plot Interaction
vis.gam(gam.7, view=c("chal.spend.raw",
  "checks.raw"), theta=325, se=FALSE,
xlab="Number of Overdrafts", ylab="Challenger
  Spending", color="bw",
plot.type="persp", zlim=range(seq(20,60, by=10)),
  type="response", ticktype="simple")
```

Example of a Poisson GAM from Chapter 6 and Figure 6.7

```
library(mgcv)
#Parametric Model
mod.1 <- gam(nulls~ tenure + congress + unified,
  data=Scourt, family=poisson) summary(mod.1)
#GAM
mod.2 <- gam(nulls~ tenure + s(congress, bs="cr")
  + unified, data=Scourt, family=poisson)
summary(mod.2)
#Compare Fits
anova(mod.1, mod.2, test="Chisq")
#Plot Result
plot(mod.2, rug=FALSE, ylab="Propensity to Overturn
  Congressional Acts", xlab="Congress", shift=-3.31)
```

Bibliography

Akaike, H. 1973. "Information Theory and an Extension of the Maximum Likelihood Principle." In *Second International Symposium on Information Theory*, ed. B. N. Petrov and F Cszaki. Akademiai Kiado.

Bartels, Larry, and John Zaller. 2001. "Presidential Vote Models: A Recount." *PS: Political Science & Politics* 34: 9–20.

Beck, Nathaniel, Jonathan N. Katz, and Richard Tucker. 1998. "Taking Time Seriously: Time-Series-Cross-Section Analysis with a Binary Dependent Variable." *American Journal of Political Science* 42 (October): 1260–1288.

Beck, Nathaniel, and Simon Jackman. 1998. "Beyond Linearity by Default: Generalized Additive Models." *American Journal of Political Science* 42 (April): 596–627.

Bellman, R.E. 1961. *Adaptive Control Processes*. Princeton, N.J.: Princeton University Press.

Berk, Richard A. 2006. *Regression Analysis: A Constructive Critique*. Thousand Oaks, CA: Sage Publications.

Bolzendahl, Catherine I., and Daniel J. Myers. 2004. "Feminist Attitudes and Support for Gender Equallity: Opinion Change in Women and Men, 1974–1998." *Social Forces* 83 (December): 759–790.

Box-Steffensmeier, Janet M., and Bradford S. Jones. 2004. *Event History Modeling: A Guide for Social Scientists*. New York: Cambridge University Press.

Box-Steffensmeier, Janet M., and Christopher J. W. Zorn. 2001. "Duration Models and Proportional Hazards in Political Science." *American Journal of Political Science* 45 (October): 972–988.

Cameron, A. Colin, and Pravin K. Trivedi. 1998. *Regression Analysis of Count Data*. New York, NY: Cambridge University Press.

Cameron, A. Colin, and Pravin K. Trivedi. 2005. *Microeconometrics: Methods and Applications*. New York, NY: Cambridge University Press.

Casella, George, and Roger L. Berger. 2002. *Statistical Inference*. 2nd edn. Pacific Grove, CA: Duxbury.

Cleveland, William S. 1979. "Robust Locally Weigthed Regression and Smoothing Scatterplots." *Journal of the American Statistical Association* 74 (368): 829–836.

Cleveland, William S. 1993. *Visualizing Data*. Hobart Press.

Congdon, Peter. 2003. *Applied Bayesian Modelling*. Hoboken, NJ: Wiley and Sons.

Crainiceanu, Ciprian M., David Ruppert, and M.P. Wand. 2005. "Bayesian Analysis for Penalized Spline Regression Using WinBugs." *Journal of Statistical Software* 14 (September) 1–24.

Craven, P., and G. Wahba. 1979. "Smoothing Noisy Data with Spline Functions." *Numerische Mathematik* 31: 377–403.

Dahl, Robert A. 1957. "Decision-Making in a Democracy: The Supreme Court as a National Policy-Maker." *Journal of Public Law* 6 (Spring): 279–295.

Davison, A.C., and D.V. Hinkley. 1997. *Bootstrap Methods and their Application*. New York: Cambridge University Press.

de Boor. 1978. *A Practical Guide to Splines*. New York: Springer.

Deb, P., and Pravin K. Trivedi. 2002. "The Structure of Demand for Medical Care: Latent Class versus Two-Part Models." *Journal of Health Economics* 21 (July): 601–625.

Dehejia, Rajeev, and Sadek Wahba. 1999. "Causal Effects in Non-Experimental Studies: Re-Evaluating the Evalulation of Training Programs." *Journal of the American Statistical Association* 94 (December): 1053–1062.

Demaris, A, M.L. Benson, G.L. Fox, T. Hill, and J. Van Wyk. 2003. "Distal and Proximate Factors in Domestic Violence: A Test of an Integrated Model." *Journal of Marriage and Family* 65 (August): 652–667.

Demaris, Alfred. 2004. *Regression with Social Data: Modeling Continuous and Limited Response Variables*. Hoboken, NJ: Wiley-InterScience.

Efron, Bradley, and Robert J. Tibshirani. 1993. *An Introduction to the Bootstrap*. Boca Raton: Chapman & Hall/CRC.

Eilers, Paul H.C., and Brian D. Marx. 1996. "Flexible Smoothing with B-splines and Penalties." *Statistical Science* 11: 98–102.

Fan, J. 1993. "Local Linear Regression Smoothers and their Minimax Efficiencies." *The Annals of Statistics* 21: 196–216.

Fan, J., and I. Gijbels. 1992. "Variable Bandwidth and Local Linear Regression Smoothers." *The Annals of Statistics* 20: 2008–2036.

Gelman, Andrew, and Jennifer Hill. 2006. *Data Analysis Using Regression and Multilevel/Hierarchical Models*. Cambridge: Cambridge University Press.

Gelman, Andrew, John S. Carlin, Hal S. Stern, and Donald B. Rubin. 2003. *Bayesian Data Analysis*. 2nd edn. Boca Raton, FL: Chapman and Hall.

Gill, Jeff. 2002. *Bayesian Methods: A Social and Behavioral Sciences Approach*. Boca Raton, FL: Chapman & Hall/CRC.

Green, P.J, and B.W. Silverman. 1994. *Nonparametric Regression and Generalized Linear Models*. Boca Raton, FL: Chapman & Hall.

Gu, G. 2002. *Smoothing Spline ANOVA Models*. New York: Springer.

Hardin, James W., and Jospeh M. Hilbe. 2007. *Generalized Linear Models and Extensions*. 2nd edn. College Station, TX: Stata Press.

Härdle, W., P. Hall, and J.S. Marron. 1998. "How Far are Automatically Chosen Regression Smoothing Parameters from their Optimum." *Journal of the American Statistical Association* 83 (March): 86–95.

Hastie, T.J., and R.J. Tibshirani. 1990. *Generalized Additive Models*. London: Chapman and Hall.

Hastie, Trevor, Robert Tibshirani, and Jerome Friedman. 2003. *The Elements of Statistical Learning: Data Mining, Inference and Prediction*. 3rd edn. New York, NY: Springer-Verlag.

Jackman, Simon. 2000. "Estimation and Inference via Bayesian Simulation: An Introduction to Markov Chain Monte Carlo." *American Journal of Political Science* 44 (April): 369–398.

Jacobson, Gary C., and Michael Dimock. 1994. "Checking Out: The Effects of Bank Overdrafts on the 1992 House Elections." *American Journal of Political Science* 38 (August): 601–624.

Keele, Luke. 2006. "Covariate Functional Form in Cox Models." Working Paper.

Keele, Luke J., Brian Fogarty, and James A. Stimson. 2004. "The Impact of Presidential Visits in the 2002 Congressional Elections." *PS: Political Science & Politics* 34 (December): 971–986.

Li, Quan, and Rafael Reuveny. 2006. "Democracy and Environmental Degradation." *International Studies Quarterly* 50 (December): 935–956.

Loader, Clive. 1999. *Local Regression and Likelihood*. New York: Springer.

Long, J. Scott. 1997. *Regression Models For Categorical and Limited Dependent Variables*. Thousand Oaks, CA: Sage.

Maindonald, John, and John Braun. 2007. *Data Analysis and Graphics Using R: An Example-Based Approach*. 2nd edn. New York, NY: Cambridge University Press.

Maoz, Zeev, and Bruce Russett. 1992. "Alliance, Contiguity, Wealth, and Political Stability: Is the Lack of Conflict Among Democracies A Statistical Artifact?" *International Interactions* 17 (July–Sept): 245–267.

Maoz, Zeev, and Bruce Russett. 1993. "Normative and Structural Causes of Democratic Peace, 1946–1986." *American Political Science Review* 87 (September): 639–656.

McCullagh, P., and J.A. Nelder. 1989. *Generalized Linear Models*. 2nd edn. Boca Raton: Chapman & Hall/CRC.

Myers, Daniel J. 1997. "Racial Rioting in the 1960s: An Event History Analysis of Local Conditions." *American Sociological Review* 61 (February): 94–112.

Nagler, Jonathan. 1991. "The Effect of Registration Laws and Education on United States Voter Turnout." *American Political Science Review* 85 (December): 1393–1405.

Natarajan, R., and R. Kass. 2000. "Reference Bayesian Methods for Generalized Linear Mixed Models." *Journal of the American Statistical Association* 95 (March): 227–237.

Nychka, D. W. 1988. "Confidence Intervals for Smoothing Splines." *Journal of the American Statistical Association* 83: 1134–1143.

Oneal, John R., and Bruce Russett. 1997. "The Classial Liberals were Right: Democracy, Interdependence, and Conflict, 1950-1985." *International Studies Quarterly* 41: 267–294.

Oneal, John R., Frances H. Oneal, Zeev Maoz, and Bruce Russett. 1996. "The Liberal Peace: Interdependence, Democracy, and International Conflict, 1950–1985." *Journal of Peace Research* 33: 11–28.

Pinero, Jose C., and Douglas C. Bates. 2000. *Mixed-Effects Models in S and S-Plus.* New York: Springer-Verlag.

Raudenbush, Stephen W., and Anthony Bryk. 2002. *Hierarchical Linear Models: Applications and Data Analysis Methods.* Thousand Oaks, CA: Sage.

Reed, William. 2000. "A Unified Statistical Model of Conflict and Escalation." *American Journal of Political Science* 44 (January): 84–93.

Rosenbaum, Paul R., and Donald B. Rubin. 1983a. "Assessing Sensitivity to an Unobserved Covariate in an Observational Study With Binary Outcome." *Journal of The Royal Statistical Society Series B* 45 (2): 212–218.

Rosenbaum, Paul R., and Donald B. Rubin. 1983b. "The Central Role of Propensity Scores in Observational Studies for Causal Effects." *Biometrika* 76 (April): 41–55.

Rosenbaum, Paul R., and Donald B. Rubin. 1984. "Reducing Bias in Observational Studies Using Subclassification on the Propensity Score." *Journal of the American Statistical Association* 79 (September): 516–524.

Rosenbaum, Paul R., and Donald B. Rubin. 1985a. "The Bias Due to Incomplete Matching." *Biometrics* 29 (March): 159–183.

Rosenbaum, Paul R., and Donald B. Rubin. 1985b. "Constructing a Control Group Using Multivariate Matched Sampling Methods." *The American Statistician* 39 (February): 33–38.

Rubin, Donald B. 2001. "Using Propensity Scores to Help Design Observational Studies: Application to the Tobacco Litigation." *Health Services and Outcomes Research Methodology* 1 (December): 169–188.

Ruppert, David, M.P. Wand, and R.J. Carroll. 2003. *Semiparametric Regression.* New York: Cambridge University Press.

Russett, Bruce. 1990. "Economic Decline, Electoral Pressure, and the Initiation of International Conflict." In *Prisoners of War?: Nation-States in the Modern Era*, ed. Charles Gochman and Ned Allan Sabrosky. Lexington, MA: Lexington Books.

Russett, Bruce. 1993. *Grasping the Democratic Peace.* Princeton, NJ: Princeton University Press.

Sekhon, Jasjeet S. 2007. "Multivariate and Propensity Score Matching Software with Automated Balance Optimization: The Matching Package for R." *Journal of Statistical Software* Forthcoming.

Silverman, B.W. 1985. "Some Aspects of the Spline Smoothing Approach to Nonparametric Regression Curve Fitting." *Journal of the Royal Statistical Society, Series B* 47: 1–53.

Stone, C.J. 1986. "Comment: Generalized Additive Models." *Statistical Science* 2: 312–314.

Stone, C.J., M.H. Hansen, C. Kooperberg, and Y.K. Truong. 1997. "Polynomial Splines and their Tensor Products in Extended Linear Modeling." *Annals of Statistics* 25: 1371–1425.

Sweet, J.A., L.L. Bumpass, and V. Call. 1988. *The Design and Content of the National Survey of Families and Households*. Madison, WI.: University of Wisconsin, Center for Demography and Ecology.

Therneau, Terry M., and Patricia M. Grambsch. 2000. *Modeling Survival Data: Extending The Cox Model*. New York: Springer-Verlag.

Venables, William N., and Brian D. Ripley. 2002. *Modern Applied Statistics with S*. 4th edn. New York, NY: Springer-Verlag.

Wahba, G. 1983. "Bayesian Confidence Intervals for the Cross Validated Smoothing Spline." *Journal of the Royal Statistical Society, Series B* 45: 133–150.

Wahba, G. 1990. *Spline Models for Observational Data*. Philadelphia: SIAM.

Weisberg, Sanford. 2005. *Applied Linear Regression*. 3rd edn. Hoboken, NJ: Wiley-InterScience.

Winship, Christopher, and Stephen Morgan. 1999. "The Estimation of Causal Effects From Observational Data." *Annual Review of Sociology* 27: 659–707.

Wood, Simon. 2006. *Generalized Additive Models: An Introduction With R*. Boca Raton: Chapman & Hall/CRC.

Yatchew, A., and Z. Griliches. 1985. "Specification Error in Probit Models." *Review of Economics and Statistics* 18: 134–139.

Author's Index

Subject Index

χ^2 distribution, 163, 192

additive model, 150, 165
additivity, 151
AIC, 10
    knot selection, 86, 87, 130

backfitting algorithm, 155, 158, 190
balance test, 230, 233
bandwidth, 31, 41
basis, 74–75
    cubic, 78
Bayesian estimation, 221
BC_a confidence intervals, 241, 250
best linear unbiased predictor, 100
bias-corrected, accelerated
        confidence intervals, *see*
        BC_a confidence intervals
bias-variance tradeoff, 46, 145
binning, 23
BLUP, 100
bootstrap, 236
    confidence intervals, 240
    nonlinearity tests, 254

nonparametric regression, 251
pairs, 247
regression models, 246
residuals, 247
standard error, 239

causal inference, 227
confidence bands, 129, 137, 160, 191
    Bayesian, 107, 125, 223
    bias-adjusted, 107
    local polynomial regression, 55
    loess, 55
    lowess, 55
    splines, 106
Cox model, 208
Cox regression, *see* Cox model
cross-validation (CV), 118, 124
curse of dimensionality, 65, 151

degrees of freedom
    residual, 161, 192
density plot, 201
derivative plot, 111
diagnostic plots, 142

Printed and bound by CPI Group (UK) Ltd, Croydon, CR0 4YY

27/10/2024

14580285-0001

.